"*A Little Book for New Scientists* is certainly the most concise and helpful book for young persons (and their parents) on a science career that I have ever seen. I would hope that every young person considering a career in the sciences will read this book and then take it down off their shelf about once a year—rereading it regularly as they proceed through graduate school and their post-doc years. If I'd had this book when I was beginning my career, I would have wanted it in a prominent place on my bookshelf—right next to my Bible."

Darrel R. Falk, emeritus professor of biology at Point Loma Nazarene University, author of *Coming to Peace with Science*

"This book is essential reading for Christians contemplating a vocation in science. It will, in addition, be fascinating for those who are not scientists but are interested in how and why science works the way it does. The authors not only successfully debunk common myths about the history of science and religion, they also provide a very honest and insightful account of scientific practices, including both its temptations and achievements. Too often the science-and-religion debate has resisted engaging with science studies alongside the specific religious quandaries opened up by scientific knowledge. This little book manages to achieve a great deal in just a few pages, which makes it particularly useful for introductory teaching contexts."

Celia Deane-Drummond, professor of theology, director, Center for Theology, Science and Human Flourishing, University of Notre Dame

"This is a refreshing little book about science and the Christian faith. It is not on the front lines of specific confrontations between Christians and various scientific theories; rather, it steps back and orients us to the kind of thing that science is and how that fits within the outlook of faith. Budding scientists should read it as a way of preparation for what the authors understand to be a holy career—a vocation in which one can truly serve God. Non-scientists should read it in order to understand that the perceived threat of science to Christian faith has been largely due to an imagined bogeyman rather than to accurate views about science. Reeves and Donaldson are to be commended for their service to the church and to the science-and-religion community for this clear and helpful book."

J. B. Stump, senior editor, BioLogos

"This volume is a nice resource for science-oriented students, newly believing Christians in science fields, or anyone seeking integration. The book draws us into a consistent, balanced, and active role as bridge between the realms of science and faith."

Jeffrey Greenberg, Wheaton College

A LITTLE BOOK FOR

NEW SCIENTISTS

WHY AND
HOW TO
STUDY
SCIENCE

JOSH A. REEVES &
STEVE DONALDSON

IVP Academic
An imprint of InterVarsity Press
Downers Grove, Illinois

InterVarsity Press
P.O. Box 1400, Downers Grove, IL 60515-1426
ivpress.com
email@ivpress.com

InterVarsity Press® is the book-publishing division of InterVarsity Christian Fellowship/ USA®, a movement of students and faculty active on campus at hundreds of universities, colleges and schools of nursing in the United States of America, and a member movement of the International Fellowship of Evangelical Students. For information about local and regional activities, visit intervarsity.org.

All Scripture quotations, unless otherwise indicated, are taken from THE HOLY BIBLE, NEW INTERNATIONAL VERSION®, NIV® Copyright © 1973, 1978, 1984, 2011 by Biblica, Inc.™ Used by permission. All rights reserved worldwide.

Cover design: Cindy Kiple
Interior design: Beth McGill

ISBN 978-0-8308-5144-7 (print)
ISBN 978-0-8308-9350-8 (digital)

Printed in the United States of America ♾

Library of Congress Cataloging-in-Publication Data

A catalog record for this book is available from the Library of Congress.

P	18	17	16	15	14	13	12	11	10	9	8	7	6	5	4	3	2	1
Y	31	30	29	28	27	26	25	24	23	22	21	20	19	18	17	16

*This book is dedicated
to Dwight and Linda Reeves for their
unwavering love and encouragement,
and to Frank and Patti Donaldson and to
all those who are committed to making the
mind a full partner with the heart, soul and
strength in loving and serving God.*

■ ■ ■

CONTENTS

ACKNOWLEDGMENTS

We wish to thank our students as well as our colleagues in the Samford University Center for Science and Religion for the many fruitful discussions that help form the context for this book. Our appreciation also goes out to David Congdon and our friends at InterVarsity Press for their support and assistance. Special thanks to Dave Nelson, who first gave us the idea to write a book for Christians studying science. Most of all we are grateful to our families, who selflessly granted us the time and space to work on this project.

INTRODUCTION

■■■

SURPRISES COME IN ALL SHAPES AND SIZES. It is one thing to learn that one has won the lottery and quite another to discover that one has cancer. Ironically, good as well as bad surprises can provide their own unique challenges. The scientist, for example, may find that an unexpected flash of insight is accompanied by significant expenditures of time and energy just to flesh out that moment of illumination. Those efforts, however, may themselves prove to be surprisingly meaningful and rewarding, only to be unpredictably ignored or even rejected by the scientist's peers.

For the Christian scientist, the range of potential surprises is even greater than it is for the non-Christian. The prospects for finding meaning and purpose in a scientific career conducted in a Christian context are deeper than those to be found on a strictly secular level, but trials beyond those normally expected in an already challenging field will also arise. And the challenges and opportunities are not solely personal. As we'll see, the Christian scientist has the chance to make a significant impact on the lives of others both in and outside of the Christian community. Consequently, pursuing a scientific profession as a

Christian opens unique doors not only for personal growth but for mission and ministry that enable those individuals involved to reach beyond themselves as their work becomes a calling and not merely a job.

> I am persuaded that in our time the battle between the powers of good and evil is pitched in man's mind even more than in his heart, since it is known that the latter will ultimately follow the former.
>
> Owen Barfield, *Saving the Appearances*

Certainly it can be stressful to be a scientist. To succeed one has to learn a large amount of technical material and experimental techniques while also undertaking new lines of research that have the possibility of failure. The genius of science is that it requires one to focus on a small slice of reality, dividing nature into a set of manageable problems. But as a consequence it takes many years of training and dedication before someone can make a contribution to his or her discipline.

However, it can be even more stressful to be a *Christian* in the sciences. For over a century there have been heated debates about the relationship between science and religion, debates that might make one feel pressure to choose between science and Christian faith. Some Christians are skeptical of or even outright hostile to a scientific career, saying that science is nothing more than the false "wisdom of the world" that the apostle Paul warned against (1 Cor 1:20-21; 3:19). On the other side, those in the profession are likely to encounter peers who are just as eager to evangelize for atheism as any evangelical Christian is for Christianity. Upon exposure to ideas that appear threatening, Christians in the sciences may find themselves experiencing an intellectual crisis. Or at the very least they may have trouble making connections between their everyday work and faith, losing track of the overall significance of their vocation by getting lost in the details of their research program.

The primary purpose of this book, then, is to help Christians studying and practicing in the sciences to connect their vocation with their Christian faith. At the same time the book can be useful for helping non-scientists relate to their Christian friends, relatives, fellow church members and other acquaintances who are scientists. Ministers will find it a helpful guide for connecting to the scientists in their congregations as well as to those who are prospective members. Finally, scientists who are not Christians might find it beneficial for better understanding their associates who are.

This interaction of the scientific and the religious holds numerous promises and perils. In an age when the two arenas are thought by many to be independent and even adversarial, the Christian scientist is in an ideal position to contest such naive views, particularly in regard to science and Christianity. Attempting to do so, however, will sometimes make it seem that one has abandoned certain cherished positions traditionally held by either community. This problem—which can be personal as well as social—arises in large part because of a failure of both parties to recognize that scientific and theological enterprises should each be as much about truth seeking as they are about truth protecting. Unfortunately, the former often takes a backseat to the latter as people abandon a genuine quest for truth in favor of a defensive posture to secure what might under the light of more rigorous scrutiny be found wanting. One of the dangers of much that passes for apologetics is precisely this—that it ends up being more about defending what one already believes than about having good reasons for those beliefs. The critical question is whether any belief can stand up to scrutiny.

> Science is not threatened by God; it is enhanced. God is most certainly not threatened by science; He made it all possible.
>
> Francis Collins, *The Language of God*

Although science has sharpened the arrows of philosophy and theology that have historically been aimed at conventional Christian beliefs, it is a mistake to think that those beliefs must therefore die as a result.[1] Surely the ultimately defenseless ones will—and should—succumb, but genuine truth has little to fear from such attacks. Indeed, it is only through those confrontations that truth can be discovered. As philosopher of science Thomas Kuhn has noted, "The prelude to much discovery and to all novel theory is not ignorance, but the recognition that something has gone wrong with existing knowledge and beliefs."[2]

---

[1]As historians would be happy to affirm, many of those beliefs were under attack from the very origins of Christianity. With the passage of time some of the arguments have become more sophisticated and perhaps even more numerous (consider, for example, the perceived threats from so-called higher criticism of biblical texts).

[2]Thomas S. Kuhn, "The Essential Tension: Tradition and Innovation in Scientific Research," in *The Third University of Utah Research Confer-*

Sometimes there is nothing wrong with the beliefs, but how is one to know if they are simply exempted from inspection?

Christian scientists who are willing to brave this territory will do so recognizing that this is not about outgrowing God but growing toward a fuller understanding of him (and his creation). The task will entail an appropriate mixture of logic and love. It is especially critical in an age when numerous individuals are leaving the church and others refuse to consider the potential reasonableness of Christianity in the first place.[3] One of the primary reasons for such behaviors and attitudes is a marginalization of those with genuine questions about how Christian teaching can in any way be compatible with the theories of modern science. It is at this juncture that the Christian scientist who appreciates the nature of the truth-seeking venture finds a unique ministry opportunity.

> There is no reason to suppose that arriving at truth would impoverish experience, however it might change the ways in which our gifts and energies are deployed.
>
> Marilynne Robinson, *Absence of Mind*

ence on the Identification of Scientific Talent, ed. C. W. Taylor (Salt Lake City: University of Utah Press, 1959), 171. Kuhn is speaking here about "mature sciences," but his insight generalizes to other domains as well, including religion.

[3]See The Barna Group, "Most Twentysomethings Put Christianity on the Shelf Following Spiritually Active Teen Years," Barna Group, September 11, 2006, www.barna.org/barna-update/millennials/147-most-twenty -somethings-put-christianity-on-the-shelf-following-spiritually-active -teen-years#.V14zEdQgvDc.

Of course, it would be easy to argue (and people do it all the time) that the issue here is not a failure to seek truth but the concept of truth itself. For these individuals, any expectation of finding truth is merely a delusion because truth is too nebulous to discover and therefore cannot really exist. Like Pontius Pilate, they hide behind the question "What is truth?" (Jn 18:38). But even if one adopts the tenuous position that the truth can never be known, it is hard to conceive of being unable to approach (or retreat from) some standard of accuracy or correctness about things—which then becomes what we might as well call a kind of truth.[4]

This gradual stepping toward the truth is characteristic of both Christianity and science. Historically, neither has been characterized by unified and unchanging perceptions of ultimate truth. The record of Jewish and Christian traditions, for example, reveals growth in conceptions of both God and humanity, and science, for its part, has stabilized longest on now abandoned or modified ideas. And although one might expect the current theories to last, there is reason to question whether they will.[5] Unfortunately, both the individuals who believe in God and those who don't sometimes act as though they are under compulsion to accept or reject the party line uncritically. This renders them largely incompetent to justify their beliefs, and any evidence they

---

[4]Karl Giberson recounts his frustration with getting someone who was promoting the impossibility of knowing any objective scientific truth to simply acknowledge that we could indeed accept some things (such as a round earth) as objectively true. Giberson, *The Wonder of the Universe* (Downers Grove, IL: InterVarsity Press, 2012), 197.

[5]C. S. Lewis, *The Discarded Image: An Introduction to Medieval and Renaissance Literature* (Cambridge: Cambridge University Press, 1964), 221.

do manage to produce makes them look like members of a bucket brigade, merely passing along what has been handed to them without bothering to check whether their bucket is empty or merely filled with platitudes.[6]

The journey on which the Christian scientist embarks, then, was well characterized many years ago by Francis Bacon, one of the forefathers of modern science, who stated, "In obedience to the everlasting love of truth, I have committed myself to the uncertainties and difficulties and solitudes of the ways, and relying on the divine assistance, have upheld my mind both against the shocks and embattled ranks of opinion, and against my own private and inward hesitations and scruples."[7] Omit the phrase "relying on the divine assistance" and this sounds little different from the sentiments of any other scientist. But the Christian scientist is not meant to be any other scientist, and those words make all the difference.[8]

---

[6]One occasionally hears calls for young people to become scientists in order that they might embrace a particular perspective held by the person issuing the call (e.g., to support a specific religious or secular position). This is bizarre because it misunderstands or ignores what should be the truth-seeking character of both science and religion (i.e., the scientists and Christians one should want are those who let the evidence take them forward).

[7]Francis Bacon, *The Great Instauration*, in *The Works*, vol. 8, trans. James Spedding et al. (Boston: Taggard and Thompson, 1863), 33.

[8]Compare 1 Corinthians 4:7, "For who makes you different from anyone else?"

PART ONE

WHY STUDY SCIENCE?

1

GOD AND THE BOOK
OF NATURE

*For the whole sensible world is like a kind
of book written by the finger of God.*

Hugh of St. Victor

■ ■ ■

WHY STUDY SCIENCE? One can answer this question from
different angles. One can ask why you as an individual should
study science. Doubtless, as a reader of this chapter you have
answered that question for yourself. Perhaps the natural world
has always fascinated you; maybe an influential parent or teacher
inspired you; maybe you are just good at science and so you
pursue it as a career to challenge yourself or support your family.

One can ask why we as a society should study science. The
most obvious answer is that science has improved our lives. We
live longer, healthier, more productive lives because of the tech-
nology that comes from scientific discovery. There are valid
worries about science's power to destroy human life or whether
it can satisfy the human desire for happiness. But most people

take an optimistic view that science has been a significantly positive development in human history.

Is it possible, however, to give Christian reasons for studying science? Are there particular reasons why Christians might be motivated to enter a scientific career? Spend time in Christian communities and you may hear worries about a scientific education. Perhaps a focus on scientific knowledge will distract you from spiritual matters, or scientific knowledge will puff up your pride, making it impossible to obtain wisdom. Or, more fundamentally, science will teach you beliefs that contradict what God teaches in Scripture. We address these worries in other places in this book, but this chapter focuses on the question, Are there positive Christian reasons for studying science?

We argue yes, and will explain why using the most common metaphor for thinking about theology and science in the history of Christianity: God has spoken in the books of nature and Scripture. For over 1500 years, Christians have used the metaphor of God's two books to suggest the complementarity of natural and supernatural knowledge.[1] The rest of this chapter will outline some theological lessons implicit in the metaphor.

GOD'S TWO BOOKS

The first implication of the two books metaphor is that some knowledge of God can be gleaned from nature. We might look at the night sky and be overwhelmed by the power and wisdom of God. We can grasp the intricacies of the cell and feel awed by

---

[1]Giuseppe Tanzella-Nitti, "The Two Books Prior to the Scientific Revolution," *Perspectives on Science and Christian Faith* 57, no. 3 (2005): 235-48.

the complexity of the biological machinery that sustains life. Or we might recognize the amount of time used to form creation and marvel at the patience and infinitude of God. Countless other examples could be given, ranging from intricacies of sub-atomic matter to the vastness of the universe. Christians affirm that the natural world is governed by the wisdom of God, and so science allows us to glimpse God's wisdom more fully.

> It is the divine page that you must lis-ten to; it is the book of the universe that you must observe. The pages of Scripture can only be read by those who know how to read and write, while everyone, even the illiterate, can read the book of the universe.
>
> Augustine, *Exposition of Psalm 45*

Spiritual knowledge that comes from reflecting on God's cre-ation led some early luminaries of the Scientific Revolution to argue that science is a spiritual activity. Robert Boyle, a leading figure of the Scientific Revolution and discoverer of Boyle's law in chemistry, described scientists (who were then called natural philosophers) as "priests of nature" because they were uncov-ering God's fingerprints in creation.[2] He even argued that science should be seen as a form of worship, and thus an activity suitable for Sundays. Even if today we do not want to go as far as Boyle, we can acknowledge the special thrill for Christians in being able

---

[2]Edward B. Davis, "Robert Boyle's Religious Life, Attitudes, and Voca-tion," *Science & Christian Belief* 19 (2007): 136.

to unpack the structure of the natural world. Theological beliefs we bring to science help us to understand the significance of what is being discovered.

Another implication is that since God is the author of both books, we as readers should not expect to find discrepancies between them. If we find places where nature and Scripture disagree, then it is a mistake of the readers—we simply have not read one or both of the texts correctly. The great early-church theologian Augustine said that whenever we have a sure result of science that conflicts with the Bible, the interpreter must bring the two back into alignment.[3] Thus learning about the natural world can help us interpret the Bible better. For example, until the invention of telescopes many Christians interpreted certain verses in Scripture to mean the earth was stationary. Psalm 96:10 states, "Say among the nations, 'The LORD reigns.' The world is firmly established, it cannot be moved; he will judge the peoples with equity." With the benefit of hindsight and scientific knowledge, Christians have no problem affirming that Christian Scripture does not teach the earth is stationary, though it may have been assumed by biblical writers. Science helped us to avoid an error in biblical interpretation.

If Christians do not recognize the value of science for interpreting the Bible, they might damage the credibility of Christianity by insisting that outdated science must be true in order to save traditional interpretations. Augustine also identified this error, writing:

---

[3]Augustine, *On Genesis: A Refutation of the Manichees, Unfinished Literal Commentary on Genesis, The Literal Meaning of Genesis* (Hyde Park, NY: New City, 2004).

Often a non-Christian knows something about the earth, the heavens, and the other parts of the world, about the motions and orbits of the stars and even their sizes and distances . . . and this knowledge he holds with certainty from reason and experience. It is thus offensive and disgraceful for an unbeliever to hear a Christian talk nonsense about such things, claiming that what he is saying is based in Scripture.[4]

Non-Christians have the God-given gifts of reason and experience, which can be used to understand the natural world. All truth is God's truth, and so Christians should not fear what science discovers about creation.

> We were made in the image of the Creator; we have the mind and reason to perfect our nature, and through them we have knowledge of God. And perceiving the beauties of nature carefully, we thereby recognize, as if through letters, God's great providence and wisdom concerning all things.
>
> St. Basil of Caesarea, *Homily on Thanksgiving*

Sometimes one hears the claim that true science should begin with Christian assumptions, thus creating a "Christian science" that differs from its secular counterpart. But Augustine argues that this strong skepticism of scientific inquiry could injure the faith: if Christians cannot be trusted on what can be empirically verified, then how can they be trusted on spiritual matters? Augustine himself left a rival religion for Christianity

---

[4]Ibid., 186.

after he found its leader proffering bad science, saying, "It was providential that this man talked so much about scientific subjects, and got it wrong."[5] A better position is to affirm that secular scientists may not be wrong when they make empirical claims (i.e., inference drawn from reason and experience); they just fail to see the true spiritual significance of what they study. In other words, they do not comprehend the spiritual realities to which the physical realm bears witness, with the result that secular scientists are often wrong when they try to construct a worldview based on science.

Though Scripture contains everything necessary for salvation, it is not an encyclopedia of all possible knowledge. Christians sometimes speak of Scripture as if it contained modern scientific theories or hidden knowledge of nature if interpreted correctly. Of course, it was possible for God, as Creator of the universe, to give us a Bible like this. But Christian theologians since almost the start of church history have recognized that God's revelation has been accommodated to the understanding of the cultures in which it was written, which includes beliefs about the natural

> The church is called in every age afresh to give a coherent account of its faith, to testify to that living truth with which it has been trusted, the gospel of Christ.
>
> Trevor A. Hart, *Faith Thinking*

---

[5]Augustine, *Confessions*, trans. Maria Boulding (Hyde Park, NY: New City, 1996), 118.

world. We will talk more about this principle in our chapter on biblical interpretation.

The difficulty, of course, is determining what theories in science have been empirically verified so that we may resist those who would use the authority of science to support anti-Christian conclusions. In such cases, Christians should not surrender basic beliefs in the name of "science." Yet these worries should not undermine the basic principle: where science shows us an empirical fact that conflicts with a traditional interpretation of a biblical passage, we need to reexamine our interpretations.

LIMITS OF THE TWO BOOKS METAPHOR

Having stressed the value of the two books metaphor, we do not want to push it too far. To head off misunderstandings, we will discuss some conclusions that the metaphor cannot support.

First, some have concluded that the basic message of the two books is essentially the same, thus making one redundant. If this were the case, one could build a theology from nature alone, without need of biblical revelation. The problem with this view is that the natural world leaves out crucial theological details. There seems little way to deduce the main tenets of Christian theology—especially about God as revealed in Jesus Christ—from the study of natural objects. As Francis Bacon, the famous philosopher of the Scientific Revolution, put it almost five hundred years ago, the works of God "show the omnipotency and wisdom of the maker, but not his image."[6] Christianity is a

---

[6]Francis Bacon, *The Advancement of Learning*, book 2, ed. James Spedding et al. (New York: Houghton, Mifflin and Company, 1895), 32.

historical faith, meaning its content depends crucially on events that happened in human history, particularly in the life of Jesus and his immediate followers. There are thus fundamental limits on what one can learn about God from the natural world. One cannot construct a theology based on nature alone; but for Christians, it is appropriate and beneficial to read both texts in coordination with each other. Followers of Christ in every age have constructed "theologies of nature," which are attempts to articulate what we discover in the natural world in light of Christian belief.

A second caution: both books need to be interpreted in a manner appropriate to their content. One way to stress the differences between interpreting the natural world and interpreting Scripture is to say they were composed in different languages. As Galileo famously said, the universe "cannot be understood unless one first learns to understand the language and knows the characters in which it is written. It is written in mathematical language, and its characters are triangles, circles, and other geometrical figures; without these it is humanly impossible to understand a word of it, and one wanders around pointlessly in a dark labyrinth."[7] Mathematizing nature is an important characteristic of the scientific process, which allows scientists with different religious or philosophical beliefs to draw conclusions from the data. In contrast, as we will discuss in chapter seven, the biblical interpreter does not use mathematical or experimental techniques to

---

[7]Galileo Galilei, *The Essential Galileo*, ed. Maurice A. Finocchiaro (Indianapolis: Hackett, 2008), 183.

reveal hidden meanings, but first contextualizes the passage in its original cultural context.

> Let no man think or maintain that a man can search too far or be too well studied in the book of God's Word or in the book of God's Works, divinity or philosophy, but rather let men endeavor an endless progression of proficiency in both; only let men beware that they apply both to charity and not to swelling; to use, and not to ostentation.
>
> Francis Bacon, *Advancement of Learning*

Finally, the metaphor of God's two books does not mean they should be given equal weight in terms of importance. Christianity deals with matters of eternal importance, and one can be a faithful Christian without holding any modern scientific beliefs. Indeed, the central message of Christianity has remained the same despite dramatic changes in Western philosophies of nature—whether Platonic, Aristotelian, Newtonian or modern ones centered on quantum mechanics, string theory and so on. This is not to deny development in theological doctrine over church history, but to recognize that Christians today can affirm "Jesus is Lord" just as their predecessors have done for almost two millennia. Scientists, by contrast, have discovered new things about the world that consequently have made their body of knowledge less stable over time. These discoveries may spur new theological reflections and developments, but the core of the Christian message will persist.

In conclusion, there has been a strong emphasis in Christian history on the value of studying nature. If done with the appropriate caution, which mainly avoids trying to prove too much, then it can be edifying to the faith. This is true on a personal level, where awe at the complexity of nature can lead to worship of the Creator. This is also true for the church on the whole, where a better understanding of nature can help with our interpretation of Scripture. Of course, not all see this basic compatibility of science and Christian theology. In the next chapter we will begin to explore why this is, using the history of science.

2

CHRISTIANITY AND THE HISTORY OF SCIENCE

*For many of the natural philosophers of the seventeenth
century, science and religion—or, better, natural philosophy
and theology—were inseparable, part and parcel
of the endeavor to understand our world.*

MARGARET J. OSLER, "MYTH 10: THAT THE SCIENTIFIC
REVOLUTION LIBERATED SCIENCE FROM RELIGION"
IN *GALILEO GOES TO JAIL*

■ ■ ■

HISTORY IS NOT ESSENTIAL for doing science, which explains why the history of one's discipline is only briefly discussed in most scientific textbooks. This is true even for theories still used in modern science. What matters most about Maxwell's equations, for example, is not how or why they were discovered, but whether you can use them to solve problems in physics.[1] However, history is essential for Christians who want an answer to the question, Why should Christians study science? As we will

---

[1]This is not to say that the study of history cannot provide valuable insights into the nature of science.

also see, one answer is that you are merely following in a long line of Christians who have studied science. Pick any science from any time period and it is almost certain that you will find Christians engaged in research on the topic.

Moreover, knowing the history of science will help inoculate you against common misunderstandings about the origins and nature of science. As we will explain in this chapter, there is a prominent story in our culture about the long history of conflict between science and Christianity, which you are likely to encounter often in your scientific career. It therefore is important to have a basic grasp of the history of science; otherwise you may be liable to accept parts of this story and thus assumptions about science that do not stand up to critical scrutiny. Just as the book of Genesis sets the basic framework for the rest of the biblical narrative, this scientific "origins" story helps frame one's understanding of theology and science.

TRIUMPHALIST SCIENCE

In the minds of many scientists and non-scientists, Christianity and science have obviously struggled with each other since the start of the Scientific Revolution, traditionally dated to the publication of Copernicus's heliocentric model in the sixteenth century. This conflict story depends on what we might call the *triumphalist* image of science. On this account, science is the triumph of human rationality over other types of knowledge. This approach typically stresses the role of the scientific method as a remedy for human ignorance, holding that we should put little confidence in beliefs that cannot be verified using this method. While science may not yet know everything, science is said to be a straightforwardly objective body of knowledge.

Also central to this story is the idea that the Scientific Revolution represents a decisive break from the superstitious beliefs of the Middle Ages, the time when theologians ruled. The key emphasis of the traditional story is the rise of a new view of physics that replaced Aristotelian cosmology with a mechanical view of nature, which reached its completion in the work of Isaac Newton. Newton's theory, it is said, pictured nature as a machine, meaning matter blindly follows the laws of cause and effect.

> For all the sterling work produced by a generation of historians dismantling with forensic precision the presumptive conflict between science and religion . . . their work has made scarcely a dent on leading scientific spokesmen, never mind popular consciousness.
>
> David Livingstone, *Science and Religion*

Continuing the traditional story, the displacement of Christianity by science is directly tied to the mechanistic image of nature. The mechanistic worldview supposedly gave nature a new authority to resolve disputes, thus freeing humanity from the dogmatic and oppressive rule of the clergy. Rather than looking to religious opinion, people could now let nature arbitrate between facts and opinion. The church, threatened by the growing power of science, reacted with a vengeance, seen most clearly in the trial of Galileo. Unwilling to be persuaded by the evidence of Galileo's telescopes, the Catholic Church forced him to recant under threat of torture and then sentenced him to house arrest.

Newton too became a figure in this drama, with many classic histories of science portraying him as a "foe of irrationality" and superstition who believed that God was more Clockmaker than providential overseer.[2] Those who acknowledge Newton's extensive theological and alchemical pursuits do not see them as having a significant impact on his physics, which is why many libraries ignored his theological writings when they came on the market in the middle of the twentieth century.[3]

Fortunately for those who are both Christians and scientists, this conflict story is almost completely wrong. A large body of scholarship in the history of science over the last three decades has revealed the supportive role that Christianity played in the emergence of science. We give here some of the reasons why the conflict story is false and encourage you to supplement our brief account with books listed in the "For Further Reading" section at the end of the book.

PROBLEMS WITH THE TRIUMPHALIST STORY

One problem with the triumphalist story is its simplistic recounting of the history of science. It concentrates on the great heroes, such as Copernicus, Kepler, Galileo and Newton, because it assumes physics is the most fundamental science. But there are other valid ways to understand the changes of this

---

[2]Stephen D. Snobelen, "To Discourse of God: Isaac Newton's Heterodox Theology and His Natural Philosophy," in *Science and Dissent in England, 1688–1945*, ed. Paul Wood (Aldershot, UK: Ashgate, 2004), 39-66.
[3]Richard H. Popkin, "Plans for Publishing Newton's Religious and Alchemical Manuscripts, 1982–1998," in *Newton and Newtonianism: New Studies*, ed. James E. Force and Sarah Hutton (Dordrecht: Kluwer Academic, 2004), 15-22.

period—particularly the experimental techniques that emerged out of the magical tradition—that do not support the same science-versus-Christianity storyline. The complexity of the developments of this period should make one suspicious of any simple stories about the emergence of science and its relationship to Christianity. A key lesson of recent history of science is that these stories reflect more the assumptions of those who tell them than the historical record.[4]

Another problem with triumphalism is the number of Christians who played a key role in the beginnings of science. As the historian John Henry remarks, "Whatever the tensions between religious institutions and science, it is a matter of historical fact that many, if not all, of the leading natural philosophers of the Scientific Revolution were devout believers."[5] Nicolaus Copernicus, Johannes Kepler, Galileo Galilei, Robert Boyle, Antony van Leeuwenhoek, William Harvey, Pierre Gassendi, Andreas Vesalius and even Isaac Newton (though he didn't believe in the doctrine of the Trinity) were devout Christians.

> The close relationship between natural philosophy and theology is evident in almost every area of inquiry about the natural world during the Scientific Revolution.
>
> Margaret Osler, "Myth 10"

[4]Nicholas Jardine, "Whigs and Stories: Herbert Butterfield and the Historiography of Science," *History of Science* 41, no. 1 (2003): 125-40.

[5]John Henry, "Religion and the Scientific Revolution," in *The Cambridge Companion to Science and Religion*, ed. Peter Harrison (Cambridge: Cambridge University Press, 2010), 41.

More crucial for science, however, was the way scientists' Christian identity shaped their engagement with the natural world. Here are a few examples of the way theological assumptions influenced the development of science. Francis Bacon provided an important theological rationale for the study of nature by arguing that it would lead to an increased appreciation of God's power and glory. Science should be judged by the "good fruits" it produced, as Scripture commanded of the believer.[6] Christian presuppositions can also be detected in the advocacy of experimental approaches to natural knowledge, where persons such as Bacon and Boyle argued that the effects of original sin required a cautious, experimental approach. Instead of speculating about general principles of nature, as philosophers tended to do, it would be far more helpful to focus on what happened during particular experiments.[7] Finally, the Protestant Reformation ushered in a central emphasis on the "literal" meaning of the biblical text, which was carried over to the reading of God's other book: nature.[8] The Protestant way of reading the Bible encouraged a break with premodern natural philosophy by encouraging nonsymbolic interpretations of objects in nature. Christian assumptions about God and nature helped lay the foundations for the emergence of science, which, at least for some historians, explains why modern science began in European culture.

---

[6]Stephen Gaukroger, *Francis Bacon and the Transformation of Early-Modern Philosophy* (Cambridge: Cambridge University Press, 2001), 159.

[7]Peter Harrison, *The Fall of Man and the Foundations of Science* (Cambridge: Cambridge University Press, 2009).

[8]Peter Harrison, *The Bible, Protestantism, and the Rise of Natural Science* (Cambridge: Cambridge University Press, 2001).

Christian assumptions about nature explain why the new mechanistic philosophy of nature was not widely seen as anti-Christian. In the upheaval of post-Reformation Europe, theologians became increasingly worried about the danger of skepticism or, even worse, atheism. For thinkers such as Gassendi and Boyle, the mechanical philosophy provided a satisfactory defense for God's existence. While this is surprising—because modern expectations closely associate reductionism and atheism—"the paradox is that among those seventeenth-century scholars who did most to usher in the mechanical metaphors were those who felt that, in so doing, they were enriching rather than emasculating conceptions of divine activity."[9] According to Gassendi, the proof of God's existence is an empirical inference from the nature of matter. Because matter is inert, it does not have the ability for self-motion, much less to organize in the complex ways displayed in the natural world. Thus Christians helped to introduce and encourage a mechanistic approach to nature, a philosophy of nature that was only seen as problematic for Christianity centuries later.

From a historian's perspective, the Galileo affair turns out not to be the clear-cut Christianity-versus-science story that is often depicted. For one reason, the trial of Galileo had more to do with his ability to alienate and embarrass supporters: the pope had encouraged Galileo's publication of *Dialogue Concerning the Two Chief World Systems*, but Galileo chose to put the pope's position in the mouth of Simplicio (i.e., Simpleton),

[9]John Hedley Brooke, *Science and Religion: Some Historical Perspectives* (Cambridge: Cambridge University Press, 1991), 159.

the foolish defender of geocentric cosmology. Moreover, the Church was defending the scientific consensus of the day against Galileo—some of whose own arguments for a heliocentric universe, such as the claim that tides were caused by the earth's movement, were not correct. This is not to absolve the Church authorities of mistakes—especially judged from modern standards—but it is to say that added context changes our interpretation of the event.

> The Roman Catholic Church gave more financial and social support to the study of astronomy for over six centuries, from the recovery of ancient learning during the late Middle Ages into the Enlightenment, than any other, and, probably, all other institutions.
>
> Neil Heilbron. *The Sun in the Church*

Recent historians have also shown why Newton cannot fit in the conflict narrative. By no longer focusing exclusively on his natural philosophy and mathematics, a new generation of Newton scholars have explored the "other Newton," the person who wrote more than four million words on theology and one million on alchemy. By only focusing on his scientific works, we get a limited picture of his thinking. To make this point Richard Popkin has argued, partly tongue in cheek, that the question is not "why one of the world's greatest scientists should have spent so much time thinking and writing about religious matters," but "why did one of the greatest anti-Trinitarian theologians of the seventeenth century take time off to write works on natural science, like the *Principia Mathematica*?"[10] Some of Newton's theology was unorthodox, to be sure, but he was no Enlightenment rationalist. From Newton's intense interest in biblical prophecy to his studies of alchemy and celestial mechanics, it was his interest in how God acts in the world that gave Newton's activities a coherence that interpreters often overlook when focusing on his scientific work to the exclusion of everything else.

To conclude this subsection, Christianity and science have had a long and fruitful history, which is what you might expect given the prevalence of the "God's two books" metaphor in Christian theology. As the historian Peter Harrison argues, the "study of the historical relations between science and religion does not reveal any simple pattern at all. In so far as there is any general trend, it is that for much of the time religion has facilitated scientific endeavour and has done so in various ways."[11]

TWO VISIONS OF SCIENCE

If science and Christianity have historically had a close relationship, how did we get to our contemporary situation, where science is seen as a secular enterprise? Two related changes occurred at the end of the nineteenth century that would have important ramifications for the relationship of science and Christianity. The first is that science became professionalized, meaning standards were implemented that could distinguish scientists from amateurs.[12] Up until this period, science was

---

[10]As quoted in Snobelen, "To Discourse of God," 44.

[11]Peter Harrison, *Cambridge Companion to Science and Religion*, 4.

[12]Frank M. Turner, "The Victorian Conflict Between Science and Religion: A Professional Dimension," *Isis* 69 (1978): 359-61.

often undertaken by "gentlemen" or clergy—in other words, those with sufficient leisure time and resources to pursue their investigations—because scientific research was not a part of the educational mission of colleges. The second change was the establishment of methodological naturalism as the standard for scientific debates, meaning scientists increasingly avoided invoking the supernatural as an explanation for phenomena within the natural world.[13]

> It is not the greatest of modern scientists who feel most sure that the object, stripped of its qualitative properties and reduced to mere quantity, is wholly real. Little scientists, and little unscientific followers of science, may think so. The great minds know very well that the object, so treated, is an artificial abstraction, that something of its reality has been lost.
>
> C. S. Lewis, *Abolition of Man*

Methodological naturalism is controversial among Christians, but it need not threaten Christian belief, as illustrated by the countless number of scientists who have practiced their faith over the past 150 years. While methodological naturalism does limit the scope of scientific explanations, it also means that there are many aspects of our world that lie beyond the reach of

---

[13]Jon H. Roberts and James Turner, *The Sacred and the Secular University* (Princeton: Princeton University Press, 2000), 28-31.

science. As an analogy, we might say that scientists are naturalists in the same way that car mechanics are. No one finds it philosophically troubling when your mechanic searches for a naturalistic explanation for the odd noise coming from the engine. In the same way, we look to scientists for answers about how natural systems normally operate.

Nevertheless, for advocates of the triumphalist account of science recounted above, methodological naturalism does not go far enough. For them, the essence of science is a commitment to a fully naturalistic account of the world, a position often called *scientific naturalism* or *scientism*. Scientific naturalism leads many scientists to postulate naturalistic accounts of subjects where we still do not have a good scientific model of what is happening. For example, they might argue that those things humans find most significant in the world— such as the love of our parents or children—are nothing but blind chemical reactions in the brain. On this view, the essence of science is a commitment to a naturalistic story coupled with the ambition to offer a micromechanical explanation for everything.

The problem with scientific naturalism, however, is that it loses its accountability to experimental data, which is the essential element of scientific inquiry. Scientific naturalism is self-defeating because it explains away the conscious awareness, reasoning and values that motivate scientific inquiry in the first place. As physicist and priest John Polkinghorne argues, the fundamental lesson of science is that reality is abundantly more surprising than we are able to imagine, and so we should recognize the inherent limits on our ability to rationally deduce the way things must be.[14] We should thus not put too much stock in the meta-theories that scientific naturalists tell us about the world and ourselves. For Christians, science can help inform one's worldview, but it should not become an all-encompassing worldview because it offers only one window into a complex reality.[15] Christians should see the eternal picture to which science lacks access. We will say more about this in upcoming chapters.

[14]John Polkinghorne, *Faith, Science and Understanding* (New Haven, CT: Yale University Press, 2001), 29-30.

[15]Josh Reeves, "On the Relation Between Science and the Scientific Worldview," *The Heythrop Journal* 54, no. 4 (2013): 561.

3

SCIENCE AND ETHICS

I think you are right in speaking of the moral
foundations of science; but you cannot turn it around
and speak of the scientific foundations of morality.

ALBERT EINSTEIN

■ ■ ■

THE FIRST AND SECOND CHAPTERS offered theological and historical defenses of why Christians should study science. This chapter argues that Christians need to be involved in science for ethical reasons. Modern science is too big and powerful—both in terms of the resources it requires and the outputs it produces—for Christians to leave scientific research to others. Examples of ethical dilemmas raised by science are everywhere—from weaponry to genetic alteration to the attempt to create artificial life forms. But the ethical questions are not reserved for a special few research topics: we argue here that science itself should be seen as an ethical activity. Understanding the ethical nature of scientific inquiry is an important step in seeing how one's Christian faith and one's scientific career relate to each other.

Before arguing for the ethical nature of science, we will first argue for the "moral ordinariness" of scientists themselves. In other words, scientists as a group are not superior to their fellow citizens. Strange as it may sound to some, this has actually been a topic of debate in the history of science.

THE VIRTUOUS SCIENTIST?

There is a long history of people claiming that scientists are ethically superior to their fellow citizens. During the Scientific Revolution this was often attributed to the theological nature of science, based on the belief that nature was a book written by God. In 1775, the English chemist Joseph Priestley wrote of the scientist: "Contemplation of the works of God should give sublimity to his virtue, should expand his benevolence, extinguish every thing mean, base, and selfish in [his] nature."[1] Just as studying Scripture should improve the moral outlook of its interpreters, so the same principle holds with the study of nature.

Over time a secular reason for scientific virtue became more common: scientists were virtuous because of their possession of the scientific method. On this view, the scientific method was unique in its ability to instill intellectual honesty. It is the scientist who truly understands the limits of human reason—by combating his or her own inclination to impose false interpretations onto nature—and thus can stand against useless speculation and

---

[1] As quoted in Steven Shapin, "The Way We Trust Now: The Authority of Science and the Character of the Scientists," in Pervez Hoodbhoy, Daniel Glaser and Steven Shapin, *Trust Me, I'm a Scientist* (London: British Council, 2004), 49.

superstition. For instance, French physiologist Claude Bernard argued, "The experimenter's mind differs from the metaphysician's or the scholastic's in its modesty because experiment makes him, moment by moment, conscious of both his relative and absolute ignorance. In teaching man, experimental science results in lessening his pride more and more."[2] This theme of scientific modesty is evident in the work of a popular philosopher of science in the twentieth century, Karl Popper. For Popper, a natural scientist can never profess with certainty the truth of a theory because no amount of positive evidence can prove it. What separates science from other disciplines is its complete and utter commitment to rooting out error. As one recent historian describes Popper's position, "The scientist becomes the only truly intellectually honest person, for only the scientist is so concerned for truth that he works on the assumption that his own theories are false."[3] The scientific method, it would seem, produces knowledge and ignorance at the same time.

As we discuss in chapter five, we think Popper's account of science only partially matches how scientific communities work. But in any case, there are good reasons to be skeptical of the idea that scientists are more immune to human frailties than the rest of society. As the famous sociologist Robert Merton argued in 1942, there is "no satisfactory evidence" that scientists are "recruited from the ranks of those who exhibit an unusual degree

[2]As quoted in Lorraine Daston and Peter Galison, "The Image of Objectivity," *Representations* 40 (1992): 122.

[3]Stephen Gaukroger, *The Emergence of a Scientific Culture: Science and the Shaping of Modernity 1210–1685* (New York: Clarendon Press, 2007), 30.

of moral integrity."[4] From a Christian perspective, the study of nature can be edifying for Christians, but there is little reason to think it, on its own, can instill the fruits of the Holy Spirit. The true significance of nature will not be obvious to those who lack the "eyes to see," thus blocking any moral benefit.

> By the end of the eighteenth century a new possibility for the character of the man of science had begun to open up, although the full development of that character was not to occur for many years. The man of science might be conceived of as someone who was neither particularly godly, nor particularly virtuous, nor particularly polite. It could be considered that there was nothing very special about the sorts of people drawn to the study of the natural world, nor anything very special about the effects on character wrought by the study of the natural world.
>
> Steven Shapin, *The Image of the Man of Science*

In retrospect, the persistence of claims made for scientific virtue was tied to building the reputation of science.[5] While the image of the virtuous scientists began because of the Christian context in which science started, it persisted because it helped create trust in scientists. The public is often not able to verify the results or potential of scientific research, so they must determine whether scientists are trustworthy and worthy of investment.

---

[4]Steven Shapin, *The Scientific Life: A Moral History of a Late Modern Vocation* (Chicago: University of Chicago Press, 2009), 21.
[5]Steven Shapin, *The Scientific Revolution* (Chicago: University of Chicago Press, 1996), 94-95.

Stories about the scientific method help give the public a sense of confidence in scientific research.

The moral ordinariness of scientists does not mean that they are to be especially distrusted when compared to other groups. Our society tends to see scientists in either/or terms. Either they are especially trustworthy, playing the role of priests who can produce sacred truths for a secular society, or they are a corrupt institution because they are beholden to political pressures.[6] A better view, as the sociologist Harry Collins has explained, is to see scientists in a middle-ground category; scientists are merely experts, and as such "should be accorded all the attention and respect we give other experts in our society like potters, carpenters, real estate agents, and plumbers."[7] The main reason we trust experts is that they participate in communities that evaluate their actions. Just as an electrician must put in many hours of training and study to gain a license to work, we trust scientists because they have undergone years of training and have been judged by their peers to be qualified. More importantly, if they violate the norms that govern scientific activity— for example, by publishing incorrect data—then they can be expelled from the community. We trust individual scientists because they are part of an institution that requires members to justify their assertions to each other using rationality and standards of evidence.

---

[6]Trevor Pinch, "Does Science Studies Undermine Science? Wittgenstein, Turing and Polanyi as Precursors for Science Studies and the Science Wars," in *The One Culture? A Conversation About Science*, ed. Jay Labinger and Harry Collins (Chicago: University of Chicago Press, 2001), 22.
[7]Ibid., 25.

FACTS AND VALUES

We now return to the topic raised at the beginning of the chapter: the ethical nature of science. At some point, practicing scientists will likely encounter what is called the fact/value distinction.[8] The fact/value distinction holds that values—what one judges as being worthy of being promoted or advanced—are human projections imposed on nature, not inherent within nature itself. Consequently, the scientist can only tell you what the world is like, not, from a scientific perspective, what the world should be like. From this perspective, one should distinguish between descriptive statements that reflect the world and normative statements that reflect human desires. If our descriptions of nature include no moral premises, then we can draw no moral conclusions.

> Of all signs there is none more certain or more noble than that taken from fruits. . . . Wherefore, as in religion we are warned to show our faith by works, so in [natural] philosophy by the same rule the system should be judged of by its fruits, and pronounced frivolous if it be barren.
>
> Francis Bacon, *The New Organon*

At first glance, this distinction seems pretty reasonable. Not only does it reflect a popular view of the natural world as indifferent to

---

[8]For more on the fact/value distinction, see Josh Reeves, "Values and Science: An Argument for Why They Cannot Be Separated," *Theology and Science* 14, no. 2 (2016).

human desires, but it also sets up a division of labor. Scientists provide the facts and society decides what should be done as a consequence. Nevertheless, the fact/value distinction is not a helpful way of understanding how a scientist does science. While in theory there may be good reasons to defend the fact/value distinction as a way to promote truth, in practice it is not an accurate way to describe scientific inquiry. It problematically suggests that scientists on a day-to-day basis offer a straightforward, perspective-free account of reality. As long as scientists are doing normal research, questions of ethics or politics do not pertain to their work. But even if it is possible in principle to have a value-free description of an objective reality, scientific theories cannot be value free because of the way humans reason about the world.

For one reason, scientific theories are always simplifications of reality, otherwise they would get too unwieldy to be useful. Thus what scientists seek about the world is not just truth—there are too many truths in the world to catalog them all—but significant truths, truths important from a human perspective.[9] The simplification is easily seen when scientists draw broader conclusions about the state of their field and "science" overall, for they must choose which subset of evidence to represent. When scientists tell these larger narratives, even when accurately presenting the scientific facts, one can always question whether their values have unduly influenced their presentation. Given the vast amount of scientific research, why include these facts and why leave others out? Many of the debates of the last century in science—the role

---

[9]Philip Kitcher, *The Advancement of Science: Science Without Legend, Objectivity Without Illusions* (New York: Oxford University Press, 1995), 94.

played by our genes, the directedness of evolution, whether nature is deterministic—are helpfully analyzed against the large cultural values presumed by the particular scientist.

This is not to say that science is merely a projection of our values—empirical science can indeed make progress in dismissing hypotheses because they are unsupported. It is to say, however, that multiple ways of reading the significance of the empirical data are always possible, and values cannot be eliminated from those interpretations. Our values function as auxiliary hypotheses, indirectly influencing the beliefs formed about our direct experience. With no way to talk about the world without at the same time theorizing it from a particular viewpoint, we cannot neatly separate facts and values in our minds. Scientists should not be seen as computers following the algorithm of the scientific method, but as detectives who attempt to make good decisions about which leads are most promising. Decisions about which theories are best supported depend on expert judgment, which cannot be walled off in one's mind from ethical, metaphysical and cultural assumptions.

> Knowledge is not one-dimensional. It is not arrived at by one strategy or method. The methods of natural science, while uniquely powerful in their chosen domain, are not applicable to much of the knowledge we know.
>
> Ian Hutchinson, *Monopolizing Knowledge*

Moreover, scientists cannot escape the consequences of their research by appealing to a value-free ideal of science because each step along the path of scientific inquiry involves actions that can be ethically evaluated. Whether for applications that are seen as morally suspect (e.g., weapons research), as praiseworthy (e.g., many types of medical research) or, more likely, as neither (e.g., research into the chemical composition of stars), the acquiring of new knowledge requires the researchers to shape and be shaped by the world around them. Even when a research topic has no obvious ethical implications, scientific inquiry can be evaluated based on the research paths not taken. Since the search space that the sciences attempt to map is so vast, what questions will we attempt to answer using our finite resources? The paths scientists choose to explore and what money agencies choose to invest are expressions of what they value. Scientific activity should always be accountable to the public, not only for the consequences of the research, but also for the research problems being investigated.

The importance of values in scientific inquiry is heightened when we recognize that the structure of scientific inquiry has changed over the past century. We now live in the era of "big science." The majority of science since World War II has been funded by large companies or national governments, a shift encouraged by the Cold War. President Dwight Eisenhower noted this in his farewell address to the nation: "Today, the solitary inventor, tinkering in his shop, has been overshadowed by task forces of scientists in laboratories and testing fields. . . . The prospect of domination of the nation's scholars by Federal employment, project allocations, and the power of money is ever present—and is gravely to be regarded."[10] Eisenhower was wary of the transformation of science because governments and corporations are often interested

---

[10]Quoted in Shapin, *Scientific Life*, 166.

in projects that maximize profits or destruction, shaping scientific development in ways that do not always benefit society. In light of the tremendous power and moral responsibility of those who fund science, the fact/value distinction can be dangerous because it allows the choices of big companies and politicians to remain hidden behind a cloak of neutrality.[11]

The era of big science has consequences for Christians who wish to conduct themselves with integrity in their scientific profession. The very existence of some laboratories makes evident the often complex entanglement of facts and values in the day-to-day world of science. Many types of scientific research are too expensive to carry out alone, requiring one to enlist the support of an institution, whether a government, university or private company. Such funding sources can be a wonderful opportunity to expand one's own research program, but they also can make researchers feel pressure—explicitly or implicitly—to slant one's focus or the interpretation of results in ways that are favorable to the sponsoring institutions. Or at the very least, it can make researchers feel powerless to protest unjust or unethical decisions for fear of losing funding. In science as in other professions, there will be opportunities for courageous moral decisions that demonstrate one's commitment to Christ and his kingdom.

CONCLUSION

We conclude by drawing our discussions of scientific virtue and the fact/value distinction together. On the one hand, scientists should

---

[11]Harold Kincaid, John Dupré and Alison Wylie, eds., *Value-Free Science? Ideals and Illusions* (New York: Oxford University Press, 2007), 4.

be recognized as experts because their technical expertise is needed in order to accurately size up some of the problems facing modern societies. It is hard to know what to do about climate change, for example, without understanding the accuracy and reliability of different types of evidence for it. And yet scientists are morally ordinary persons, meaning that once they have offered their technical opinions, they have no special ethical or political expertise about how best to solve the problems they identify. As the physicist Ralph Lapp argued in the 1960s: "Scientists as a group probably have no better sense of human values than any other group. To say that science seeks the truth does not endow scientists as a group with special wisdom of what is good for society."[12] This is not to say that scientists cannot offer good solutions to certain problems, only that they should not have a monopoly on policy decisions.

> The natural understanding of the value-laden character of our world is that there is a supreme Source of Value whose nature is reflected in all that is held in being. Otherwise the pervasive presence of value is hard to understand.
>
> John Polkinghorne, *Belief in God in an Age of Science*

An advocate of the fact/value distinction would at this point recommend that the scientist stick only to the facts and leave everything else to society. The problem with this position, as described above, is the fact/value distinction offers a misleading picture of scientific judgment, which allows values to sneak in

---

[12]Shapin, *Scientific Life*, 71.

without our being aware of them. It is far better to recognize our values than to pretend they do not exist.

Rather than denying the role of values in science, we should ask instead whether any particular scientific analysis has been abnormally skewed by someone's values. The way to determine bias is by having one's scientific work evaluated by other scientists to see what theories generate consensus. Scientific consensus is not infallible; we can think of many stories of famous scientists who disagreed with their peers and were ultimately proven correct. But even in these cases, the scientists are remembered because the scientific community eventually came around to see the merit of their theories.

If scientific consensus is going to be relied on as an indicator of scientific truth, then the scientific community needs to be a broad and diverse community, instead of consisting of one particular culture or perspective. Just as we are more likely to trust a government policy decision crafted by representatives of different constituencies, so too we are more likely to trust a scientific conclusion if it is backed by scientists from various backgrounds. As we will discuss more in later chapters, if Christians abandon science for fear of anti-Christian bias in scientific research, it only ensures that no Christians will be left to evaluate claims made in the name of science.

Once the role of values is recognized in science, it becomes easier to see how one's faith can impact one's scientific career. Christians are free to pursue outcomes that align with their convictions: alleviation of suffering, cherishing of life and stewardship of the earth, for example. And Christians can object to scientific research that runs counter to these goals. In other words, one's scientific career can be motivated by the same Christian beliefs that guided our predecessors in the Scientific Revolution.

PART TWO

CHARACTERISTICS OF FAITHFUL SCIENTISTS

4

HOPE IN THE FACE
OF ADVERSITY

Consider it pure joy, my brothers and sisters, whenever
you face trials of many kinds, because you know that
the testing of your faith produces perseverance.

JAMES 1:2-3

Those who hope in the LORD will renew their strength.
They will soar on wings like eagles; they will run and
not grow weary, they will walk and not be faint.

ISAIAH 40:31

■ ■ ■

AS WE'VE SEEN, people become scientists for a variety of
reasons, not the least of which include the excitement of discovery, a chance to exercise particular skills and abilities, and
the prospects of recognition for their accomplishments. Indeed,
these are some of the more obvious potential rewards of being
a scientist, but science, like any other discipline, is not without
its hardships. Rather than being caught off guard, knowing

some of the more common sources of adversity allows the scientist to be better prepared when confronted with them. Furthermore, the Christian scientist has resources to call on that may be unavailable or more difficult to access for those who are not Christians.

MEETING PROFESSIONAL AND RELATIONAL DEMANDS

In his book *Consilience*, the eminent biologist E. O. Wilson paints a painful picture of some of the demands made on a successful scientist.[1] Depending on place and type of employment, those demands could involve any or all of the following: research, teaching, administration, reading, writing, presentations and travel. Ultimately every scientist is faced with the dilemma of trying to stay current in a world where knowledge is rapidly increasing. Most scientists will also attempt to maintain good rapport with family and colleagues, trying to be a good spouse and parent and seeking to retain some semblance of a social life. Many will want to pursue a hobby or two. A few will consider their civic responsibilities. For Christian scientists, a relationship with God and their churches plus a calling to be involved in various church-related ministries will exact a further toll on their time.

> Every gift of noble origin is breathed upon by Hope's perpetual breath.
>
> William Wordsworth, *Collected Poems*

---

[1]Edward O. Wilson, *Consilience: The Unity of Knowledge* (New York: Vintage, 1999), 60.

It seems fair to ask why anyone in his right mind would want to endure that, but as long as the rewards (such as those mentioned earlier) are deemed sufficient, the scientist will strive to make things work. It would be ridiculous, of course, to think that this dilemma is limited to the sciences. Many fields are characterized by significant job-related responsibilities and a need to master a changing body of knowledge while attempting to balance a host of competing demands on one's time, and scientists can take some comfort in the fact that they are not alone nor in the only field where such occurs. The Christian in the sciences or in any comparably demanding career, however, will be conscious of where his true priorities are at least supposed to reside, and will attempt to live accordingly in the belief that doing so is part of what God expects. On occasion this will lead to frustration when one is aware that some project could really benefit from the time that is currently being put elsewhere. However, the big-picture perspective can also be used to give special meaning not only to the nonscientific activities that require engagement but also to the scientific ones that can otherwise periodically seem stale or pointless. Believing that one is exercising God-given abilities to accomplish something that has possible value for others and that one's scientific explorations can provide potentially useful insights into big questions of life (a topic to which we will return later) offers a special type of hope for meaning and purpose that is largely unavailable to the nonreligious.

Despite the hopefulness of these distinctly Christian views of a scientific career, it is important to recognize that other positions are possible. Thus in the battle to find time for what can seem like an impossible list of demands, in the single-mindedness that can emerge as mastery of one's discipline grows, and in the knowledge that science itself may be more fun and possibly more tangibly rewarding than other areas (or at least requires what seems to be less sacrifice), there is the ever-present danger that science (or one's scientific career) becomes the new god. As supposed masters of logical reasoning, scientists will have at their disposal any number of ways to justify their behavior and suppress any guilt that ensues. They might, for instance, look with envy at some of their non-Christian colleagues who have made science a god and note that by doing so they save significant amounts of time that would otherwise be spent in traditional religious activity. Alternatively (or in conjunction with this), they could convince themselves that the fruits of their scientific labors justify shirking or ignoring other areas.

Unfortunately this is not a threat that diminishes as a career matures. The Christian scientist who is mindful of it, however, will have taken the first step toward avoiding this error. That mindfulness will occur in the context of recalling that it is no good "for someone to gain the whole world, yet forfeit their soul" (Mk 8:36), as well as seeing one's collective Christian responsibilities as a gift, rather than a burden, that is ultimately part of an abundant life (Jn 10:10).[2]

THE PROBLEM OF SPECIALIZATION

It is important to keep in mind that one of the reasons many scientists fail to have the big-picture perspective is simply a result

---

[2]This entails recognition that there is a vast difference between knowing about God (a task toward which inquisitive, analytical minds naturally gravitate) and knowing him. The former, however, may enable the latter.

of the problem of specialization. The specialist becomes an expert in a small corner of the universe and thus possesses a great amount of knowledge, but such an exclusive focus may render one unable to see the universe in broader strokes. Although single-mindedness is sometimes useful, standing too close to anything can induce a form of short-sightedness that not only inhibits engagement with larger issues but can actually become a liability in the attempt to solve the very problems on which one is focused. This is because useful perspectives on some problem at hand may be available from areas that superficially seem foreign to one's chosen field.

> We choose the methods we use to be essentially those that pick out only regularities with which we are somehow already very familiar from our own built-in powers of perception.
>
> Stephen Wolfram, *A New Kind of Science*

The physicist Per Bak has suggested, "Perhaps our ultimate understanding of scientific topics is measured in terms of our ability to generate metaphoric pictures of what is going on. Maybe understanding is coming up with metaphoric pictures."[3] If Bak is correct, then the source of the necessary metaphors— precisely because they are metaphors—must often lie outside the discipline. Recognizing this encourages the scientist to engage in

---

[3]Per Bak, *How Nature Works: The Science of Self-Organized Criticality* (New York: Springer, 1999), 50.

the habit of zooming in and out, a form of cross-training that both serves as an antidote for the myopia that can prevent progress toward solving some specific problem and helps relieve the stagnation that tends to accompany any activity that is pursued without variety.

In short, a single-mindedness that all too easily degenerates into a blinding narrow-mindedness is a real threat. However, in the context of seeing a career as a calling, coupled with biblical injunctions to be a good steward of abilities and time, the Christian scientist has additional weapons with which to counter this tendency.

LIVING WITH REJECTION

Noted writer Arthur Herman recounts the sad plight of the famous physicist Ludwig Boltzmann, who committed suicide in the wake of rejection of his kinetic theory, only to have the theory vindicated soon thereafter.[4] Now, few scientists will achieve the scientific insight of Boltzmann or be driven to such extreme measures. Nevertheless, rejection is a difficult pill to swallow no matter what form it takes, and for the scientist those forms can be many. Failed experiments, refusal to accept articles submitted for publication, lack of funding (perhaps due to failed grants), absence or withdrawal of peer support, even self-criticism engendered by a temporary lack of insight—all can contribute to a sense of rejection that may leave a scientist wondering whether she has chosen the appropriate career. Sometimes a change is

---

[4]Arthur Herman, *The Cave and the Light: Plato Versus Aristotle, and the Struggle for the Soul of Western Civilization* (New York: Random House, 2013), 477-83.

indeed in order, but more often than not the scientist must simply realize that facing rejection is not unusual, especially while trying to become established. In any of these respects Christian scientists, like any other, will never be satisfied with less than their best. Yet for the individual who stays the course, hope for improvements in productivity is reasonable. Whether those hopes play out, however, is not guaranteed. For instance, Dean Simonton points out that scientific creativity is the product not only of genius and logic but also of chance and cultural context.[5]

Thus there will always be factors outside the control of any scientist that affect both performance and reception of her work. How one responds to those factors, however, is almost always a matter of personal choice. As with any form of adversity, one's reactions will illuminate the depth of her Christian commitment with the inevitable result of strengthening or weakening her witness. As Paul noted long ago, "For Christ's sake, I delight in weaknesses, in insults, in hardships, in persecutions, in difficulties. For when I am weak, then I am strong" (2 Cor 12:10).

In addition, Christians have the promise that their worth in God's sight is not ultimately dependent on being at the top of the proverbial pyramid—that "a person's praise is not from other people, but from God" (Rom 2:29). Faith and hope that the scientist and his or her work will be meaningfully incorporated into God's bigger plan can be a tremendous source of stability and comfort despite the ups and downs that invariably accompany a scientific (or any other) career.

---

[5]See Dean Keith Simonton, *Creativity in Science: Chance, Logic, Genius, and Zeitgeist* (New York: Cambridge University Press, 2004).

> And we know that in all things God works for the good of those who love him, who have been called according to his purpose.
>
> Romans 8:28

Confronting Contentious Issues

If the history of science and religion has taught us anything it is that no scientific theory or religious conviction exists in a vacuum. As discussed in chapter nine, one of the exciting prospects for the scientist is that her scientific insights might contribute to an enhanced understanding of big questions of meaning and value. It is also true, however, and potentially scary to note that both the theoretical framework in which the scientist works and the fruits of her labors may sometimes seem to question or contradict a prevailing religious view with which she or those individuals she knows and loves have become comfortable. For Copernicus and Galileo this consisted of new views on the arrangement of the earth, sun and planets. For many scientists, at least since Darwin's day, the issue has involved attempts to reconcile theories of origins with biblical creation accounts. More recent issues include global warming, mind-brain-soul interrelationships and transhumanism. Although all scientists may be affected by these dialogs, the Christian scientist is particularly vulnerable because of the risks of losing credibility with the scientific establishment and fellowship with the religious one—or vice versa.

Seldom will an outcome be as dramatic and publicly notable as it was for Galileo,[6] but privately it can be every bit as severe. For the Christian scientist who is trying to maintain a balance in what appears to be two worlds, the easy way out may simply be to jettison participation in one or the other. That would be a sad turn of events precisely because only by scientists remaining faithful is there hope for reconciliation. Theologians and philosophers will have much to say about contentious issues at the interface of science and religion, but without participation from Christian scientists their efforts will almost always be incomplete. The scientist will be able to contribute insights that fall outside the scope of the training received by many of those in other disciplines. If scientists are not helping form responses to perceived dilemmas, the approaches that are suggested may be inadequate and should be viewed with some suspicion. Thus in all such work the Christian scientist labors in the hope that by helping to bridge the gap between scientific and theological perspectives he or she is participating in a noble search for truth and is performing a useful service to others who are engaged in the same pursuit.

THE BIG PICTURE

Hope is the attitude that considers how we would like things to be. In a very real sense it can simply be described as wishful thinking.[7] For what, then, might the Christian scientist wish? Well, a Nobel Prize would be nice! In the absence of that unlikely

---

[6]Possibly because the Christian scientist will (hopefully) show more political astuteness than Galileo did.

[7]Steve Donaldson, *Dimensions of Faith: Understanding Faith Through the Lens of Science and Religion* (Eugene, OR: Cascade, 2015), 117.

occurrence, there is nevertheless the hope that one will make a significant contribution to the discipline of choice and receive adequate recognition for doing so. This will entail making new discoveries, the hope for which is ultimately a dominant driving force for any scientist[8]—as is the hope for finding meaning and fulfillment in one's work.

But the Christian scientist is not just any scientist. For that individual, career objectives and decisions will always be seen in light of a larger context—the Christian scientist necessarily sails in deeper waters. Such individuals will be conscious of a need to ply their trade with a view toward doing God's will, furthering God's kingdom and in general demonstrating the earmarks of Christian discipleship. The Christian will understand the practical nature of a scientific career no less than the non-Christian, but will bring to their endeavors a belief that there is a God who knows them—one who has a plan for their lives and can multiply the fruit of their labors, and to whom they can turn for guidance and insight as well as correction and forgiveness.

> Coming to know Christ provides the most basic possible motive for pursuing the tasks of human learning.
>
> Mark Noll, *Jesus Christ and the Life of the Mind*

Such a belief fosters a transcendent hope that is built on faith in God's love as revealed in Christ. As with any hope it is forward looking, but it differs from the purely secular because of its

---

[8]Wilson, *Consilience*, 61.

content—it is what we hope God will do with us, for us, to us and through us. As 1 Peter 1:21 states, "Through him [i.e., Jesus] you believe in God, who raised him from the dead and glorified him, and so your faith and hope are in God."

This is the faith and hope with which Christian scientists operate. The resulting perspective—which turns their practice of science into a ministry as well as a mission—is a far cry from views such as that expressed by physicist Steven Weinberg, who believes that the more we know about the universe, the more pointless it seems.[9] This big-picture approach also frames an appropriate legacy for the Christian scientist, who could desire nothing greater on his or her tombstone than recognition of a life lived in Christian service and love.[10]

---

[9]Steven Weinberg, *The First Three Minutes: A Modern View of the Origin of the Universe* (New York: Basic, 1977), 162. But wouldn't the universe actually seem more pointless if it was incomprehensible?
[10]See John 13:12-16, 34-35.

5

LIFE TOGETHER

Working with Others in a Scientific Community

> *No man is an island entire of itself; every man is*
> *a piece of the continent, a part of the main.*
>
> JOHN DONNE, *DEVOTIONS*, MEDITATION XVII

■■■

THERE IS A LONG HISTORY of thinking the pursuit of science is best done when turning away from one's community. Francis Bacon, influential philosopher of the Scientific Revolution, famously warned against the *idols of the theater*—received wisdom that is passed on despite being incorrect.[1] Accordingly, many have taken skepticism to be the essence of scientific inquiry. Far better to be ignorant and keep searching for the truth, on this view, than to believe something false on the testimony of someone else. As discussed in chapter three, Karl Popper argued that the main job of scientists is to disprove hypotheses. Likewise, the English Royal Society adopted as a motto *Nullius in Verba* (On the Word of No One).

---

[1]Francis Bacon, *The New Organon*, ed. Lisa Jardine and Michael Silverthorne (Cambridge: Cambridge University Press, 2000), 42.

Despite the popularity of such views, the actual practice of science does not match this individualist image.[2] A community is always involved, though it may not be physically present with the scientist at every moment of time. None work in a vacuum, if for no other reason than that virtually any science of value is constructed on the foundation of previous work. No matter how skilled one becomes in a scientific career, the vastness and complexity of scientific knowledge means one cannot become proficient in all the technical details of a single discipline, much less science as a whole. There are even some situations, such as research involving the Large Hadron Collider, where one scientist cannot have full knowledge of all the parts of a single experiment. Much of scientific activity thus depends on trusting the work of colleagues and predecessors. Even the curmudgeonly Isaac Newton acknowledged, "If I have seen further it is by standing on the shoulders of giants."[3]

> The scientist takes off from the manifold observations of predecessors, and shows his intelligence, if any, by his ability to discriminate between the important and the negligible, by selecting here and there the significant steppingstones that will lead across the difficulties to new understanding.
>
> Hans Zinsser, *As I Remember Him*

---

[2]Stereotypical views of the crazed scientist working alone in his laboratory with such singleness of purpose and dedication to his task that he shuns the social graces and neglects to attend to personal needs are mostly apocryphal.

[3]Isaac Newton to Robert Hooke, February 5, 1675/6, in *The Correspondence of Isaac Newton*, ed. H. W. Turnbull, vol. 1, *1661–1675* (Cambridge: Cambridge University Press, 2008), 416.

Thomas Kuhn described the paradigmatic view of typical scientific practice—and in the process popularized the term *paradigm*—in his classic work on the interaction between "normal" and revolutionary science.[4] "Normal" science characterizes the activity of most scientists most of the time as they work in a community of other like-minded individuals. Occasionally anomalies arise that suggest the need for significant revision or replacement of an existing theory, but for the most part this is rare and (despite minor differences of opinion and interpretation) the community survives intact. The Hungarian philosopher of science Imre Lakatos extended Kuhn's view to be a bit more tolerant of anomalies in what he calls "research programs,"[5] but once again the beliefs of a professional community form the core theoretical framework, though the community itself may be spread out in both space and time.

In support of such a community the scientist will often be involved in joint experiments; contribute to coauthored papers; serve as a peer reviewer or journal editor; attend, organize and give presentations at conferences; and seek to keep current with the relevant literature. Productivity in the sciences, then, is dependent on a working interaction between scientists, even for those scientists who choose to physically work alone or are forced to by exigencies of the environments in which they labor.

---

[4]Thomas S. Kuhn, *The Structure of Scientific Revolutions* (Chicago: University of Chicago Press, 1996), and "The Essential Tension," in *The Third University of Utah Research Conference on the Identification of Scientific Talent*, ed. C. W. Taylor (Salt Lake City: University of Utah Press, 1959).

[5]Imre Lakatos, *The Methodology of Scientific Research Programmes* (New York: Cambridge University Press, 1980).

Any scientist who is attempting to push the frontiers of knowledge must interact with others.

Despite the essentially productive nature of this interaction between individuals operating within a paradigm, however, there is always the inherent danger of becoming so locked in a particular way of looking at the world and doing things that further progress is stifled. Kuhn recognized this problem, but it was the Austrian philosopher Paul Feyerabend who voiced the biggest concern, suggesting that major steps forward ultimately occurred through a type of thoughtful anarchy within scientific ranks brought on by an unwillingness to blithely accept the status quo.[6] The resulting tension, as Kuhn termed it, between normal and revolutionary science was therefore not the exception but the goal for Feyerabend, who feared that the search for truth could all too easily stagnate in an entrenched paradigm that was characterized by the absence of an ongoing comparison among competing theories.[7] Consequently, despite the communal nature of most scientific practice, there will be for a minority of scientists those times when they find themselves swimming against the current of the dominant theoretical framework as accepted and practiced by their peers. Occasionally their views will become the new norm.

This is particularly relevant for Christian scientists who, as we will discuss in detail later, may find that they are in the minority

---

[6]Paul Feyerabend, *Against Method* (New York: Verso, 1993). Note that this was to be neither a careless nor a thoughtless rebellion nor predicated on a desire to be different just for its own sake.

[7]Ibid., 22. It is important to keep in mind that the dangers of inbreeding within the sciences with which Feyerabend was concerned has its parallels in religious thinking.

regarding their religious beliefs even though they belong to the scientific mainstream. Such individuals are therefore presented with a different kind of tension in which the weight of scientific opinion can also bend their Christian perspectives beyond recognition. It is also possible, however, that for those Christian scientists working in an environment in which their dominant contacts are Christian, just the opposite can occur. In that case one's religious beliefs can distort the practice of good science or rationalize the denigration of unpopular scientific theories. It is in both of these cases that recalling the truth-seeking nature of genuine scientific and theological practice can provide a foundation from which to interact with members of one's scientific and Christian communities who may or may not at the moment share beliefs in either domain.

> We all need to build our lives on the stable, solid, and secure rock of truth rather than on the shifting sands of opinion.
>
> Alister McGrath, *Surprised by Meaning*

The humility (as described in the next chapter) that should characterize the attitudes and behaviors of the Christian scientist will not only serve as a corrective for overblown opinions about his or her own theological or scientific insights, but it will also enable communication with those whose ideas differ. Without humility, any possibility of meaningful community quickly disintegrates. Note that this doesn't mean a community must always involve uniformity of belief. Rather, what is needed is an environment in which mutual respect is fostered despite differences in belief, whether scientific or religious. Community can exist if one scientist respects the other for her humble openness to the pursuit of truth, even while not necessarily respecting her current specific opinions about a particular issue. Although the searcher for truth sometimes plays solitaire, it is usually not for long and never just for the sake of being different.[8] Believing that there really are external, beneficial sources of insight is better than thinking that one is self-sufficient because it inspires the search for such sources, encourages the corroboration of one's own insights with those of other truth seekers and helps promote the requisite humility.

Unfortunately, scientists are not always successful in achieving this type of community. Not too long before this book was written, for example, a dispute was raging between evolutionary biologists over the extent to which a long-cherished mechanism called kin selection was efficacious in explaining certain behaviors in insects and perhaps in humans.[9] The acrimony exhibited by some of the participants in that discussion belies the usual cold, calculating, entirely rational demeanor many people believe characterize scientists, but is unsurprising among people who live with

---

[8]Steve Donaldson, *Dimensions of Faith: Understanding Faith Through the Lens of Science and Religion* (Eugene, OR: Cascade, 2015), 165.

[9]For a definition of kin selection, see Edward O. Wilson, *Consilience: The Unity of Knowledge* (New York: Vintage, 1999), 183. The behavior under consideration is eusociality, defined in the paper that generated the melee: Martin A. Nowak et al., "The Evolution of Eusociality," *Nature* 466 (2011): 1057-62. For an example of controversy in cognitive science, consider the heated debate over mental imagery reflected in Zenon W. Pylyshyn, *Seeing and Visualizing (It's Not What You Think)* (Cambridge, MA: MIT Press, 2006), and Stephen M. Kosslyn et al., *The Case for Mental Imagery* (New York: Oxford University Press, 2006).

the firm belief that only their view is rational. It is particularly revealing that a number of scientists who disagreed with the new opinion (which was published in a prestigious journal) chose to vent their disapproval in a letter[10]—which suggests that attempts at enforced conformity didn't die with the Inquisition.

Despite the common image of the virtuous scientist, discussed in chapter three, scientists can often act in prideful and selfish ways. In fact, aspects of the scientific process may encourage this. As a scientist who specializes in scientific misconduct has written, "Scientists are not disinterested truth seekers; they are more like players in an intense, winner-take-all competition for scientific prestige and the resources that follow from that prestige."[11] Some scientists will be tempted to get ahead by pushing others down.

Yet scientists, especially those just embarking on their career, should avoid the temptation to treat others badly. Even from a nonreligious perspective, establishing relationships with others can play an important role in advancing one's career. Finding mentors and colleagues whose work is trusted can save research time—through feedback that helps one avoid dead ends—and can offer new opportunities. Working well with others can have a positive effect on a career.

> Show proper respect to everyone.
>
> 1 Peter 2:17

---

[10]Thomas Bartlett, "Biologists Team Up to Quash New View of Cooperation," *The Chronicle of Higher Education*, May 13, 2011, A17.

[11]David Goodstein, "Scientific Misconduct," *Academe* 88 (2002): 31.

More importantly, one should treat others well in and outside the scientific community because Christ calls us to do so. For Christian scientists who are mindful not only of their professional mission but also that they have been called to more significant tasks, learning to foster community is not just an option but a mandate.[12] The Christian scientist realizes that the relational aspects of community enable more than just a fruitful environment for scientific endeavors—they can also be crucial to any successful attempts to share the Christian message with peers. Furthermore, no one lives in just a single community. In addition to professional settings there are a variety of family, social and congregational groups to which each individual belongs, and those communities are likewise preserved as potential places of fruitful interaction only if their constituents adopt and maintain humble and respectful attitudes.

But how is community built beyond demonstrating the humility and openness already suggested? Well, a good starting point is to examine attributes that characterized the earliest Christian communities. The key idea here is that those features may also facilitate the development and maintenance of other forms of community. One finds that those Christian environments were characterized by sharing, giving, forgiving, compassionate attitudes where individuals were attuned to more than just the professional needs of community members.[13] Of course these qualities were (and should be) a direct reflection of the nature of Christ himself, who through actions and parables

---

[12]See Matthew 28:19-20; Acts 1:8.
[13]These attributes of Christian community are captured by the Greek word *koinonia*.

(e.g., the good Samaritan, Lk 10:25-37) sought to broaden perspectives on what constitutes an appropriate sense of community.

Despite the importance of community, it is almost inevitable that the Christian scientist in particular will at least on occasion feel like "the voice of one crying in the wilderness." Lest that simply degenerate into crying, there are several factors at work that can ease the potential pain. To begin, it is seldom the case that one is truly alone in his or her beliefs. If that is the case, there is probably genuine cause to worry. More often than not, however, there is still a community of likeminded Christian scientists, even if they are geographically dispersed.[14] In an age when practically everyone has telephone, email and web access, communication with such individuals is just around the electronic corner.

> Nothing we do, however virtuous, can be accomplished alone.
>
> Reinhold Niebuhr, *The Irony of American History*

Building relationships with other Christian scientists begins with discovering who they are, and that can take place in a variety of ways. Large organizations such as The American Scientific Affiliation (US) and Christians in Science (UK and beyond) provide multiple venues that enable contacts among Christian scientists. Many other organizations and conferences that are similarly aimed at melding science and Christianity

---

[14]Recall the story of the prophet Elijah who, thinking he was the only true servant of God, was ready to die, only to be informed by God that there were thousands of others who remained faithful (1 Kings 19:1-18).

offer additional sources of interaction and discussion. Reading and especially publishing in journals devoted to science and theology can also open doors to the creation of new and meaningful relationships.

It is also helpful to recall that all Christian scientists work in a virtual community of the many who have preceded them.[15] Thus, whether it be Robert Boyle finding God in his laboratory experiments, the French Jesuit paleontologist Pierre Teilhard de Chardin contemplating cosmological, biological and conscious evolution in an eschatological context, or the many who have merely gone about their (and their Father's) business in relative obscurity, Christian scientists will find a common underlying bond. Despite potential differences in views on peripheral issues, the results of such virtual encounters can be—as with those in any community—thought provoking and stimulating but ultimately supportive as well.[16]

Finally, recognizing that psychological and spiritual bonds can transcend time and place is important, but the greatest comfort for Christians, scientists or not, will come from believing that they are known and loved by God, who "will never leave you or forsake you" (Heb 13:5 NRSV; see also Deut 31:6). That is the true source of all Christian community.

---

[15]See Donaldson, *Dimensions of Faith.*
[16]They can also be non-supportive of fringe or poorly reasoned views.

6

THE KNOWN UNKNOWNS

Science and Intellectual Humility

*The confidence which men . . . have in their own
acumen is as unreasonable as the small regard
they have for the judgments of others.*

GALILEO GALILEI, *TWO CHIEF WORLD SYSTEMS*

■ ■ ■

LOOKING BACK AT SEVERAL thousand years of human
history, during which much was believed about the universe
that is now deemed to have been wrong, it is not entirely bad
form to ask, "How could the smartest Greeks have been so
dumb?" Perhaps in the not-so-distant future the same question
will be asked about today's smartest "geeks." Yet such questions
misunderstand what science does in providing more refined
views as time passes—which, by the way, is also what religion can
do (although this is lost on many individuals, religious and not,
who think religion and the theological insights that undergird it
are static enterprises). One would think that historic precedent alone should be enough to make the need for intellectual

humility obvious, yet as human knowledge explodes quite the opposite seems to occur. Rather than appreciating how individual insight as a portion of the collective body of knowledge is actually shrinking at an exponential rate, people seem no less inclined today than in the past to assume that they have arrived on the intellectual scene.

This attitude is particularly puzzling in the sciences, where the failure of certain dogmatic positions of more recent times— for example, Lorentz's views on relativity, Einstein's dismissal of quantum randomness and Hoyle's criticism of Big Bang cosmology[1]—have only partially mitigated such questionable confidence. Current unyielding positions on various evolutionary mechanisms, such as kin selection and self-organizing systems, hint that the rapid proliferation of new ideas spawned by modern science has also been accompanied by an increase in dogmatic perspectives. At the same time, a spate of relatively recent books even suggest that science itself may be running out of new things to say, implying that what scientists are now saying is the best that can be said.[2] Perhaps this is true; but any hope that the scientific mindset will prevent its practitioners from succumbing to the fallacy of ultimate insight seems oddly in danger. Worse,

---

[1]Walter Isaacson, *Einstein: His Life and Universe* (New York: Simon & Schuster, 2007), 133, 463; Karl Giberson, *The Wonder of the Universe* (Downers Grove, IL: InterVarsity Press, 2012), 76-80.

[2]For example, David Lindley, *The End of Physics: The Myth of a Unified Theory* (New York: Basic, 1994); John Horgan, *The End of Science: Facing the Limits of Knowledge in the Twilight of the Scientific Age* (Reading, MA: Addison-Wesley, 1996); Russell Stannard, *The End of Discovery: Are We Approaching the Boundaries of the Knowable?* (New York: Oxford University Press, 2010).

even scientists (as well as theologians and philosophers) who make it part of their primary mission to convey to others the need for intellectual humility—via pleas to keep an open mind and recognize one's limitations—frequently do so from an entrenched and inviolable position that has somehow escaped their own criticism.

> Do not think of yourself more highly than you ought, but rather think of yourself with sober judgment.
>
> Romans 12:3

REASONS FOR A LACK OF INTELLECTUAL HUMILITY

What then is the source of a lack of intellectual humility? How do otherwise smart people fail to acknowledge the potential for their own shortcomings? Considering the tentative history of science, why aren't scientists among the most humble of folks?

Success of science. The general success of science has made being humble problematic for some who take pride in their membership in such an intellectual and productive arena. Of course no scientist has mastered more than a smidgen of science as a whole, which means such pride is somewhat akin to spectators being proud of the performance of their favorite sports team. Surely the best any scientist can manage is to be a spectator for most of science, and there is little honor for spectators.

Personal achievement. On the other hand, it is also possible to take pride in having mastered a subject that few others have, or to revel in the products of one's labors. Scientists tend to be smart people, and intellect and humility, like oil and water, mix poorly for many. Research success, publications, speaking engagements and the like can all work against the cultivation of a humble spirit. In this regard it is convenient to ignore the fact that many individuals probably could have performed comparably had they chosen one's particular field to study rather than some other. As any college professor can attest, there are hosts of students who could excel at any number of disciplines, although each must ultimately select one or two on which to focus. Of course one might merely believe her discipline to be the most difficult and opaque to all but a very few. Yet even if that were the case, it is hard (as we are about to see) to explain why pride is justified outside one's disciplinary specialty.

> A proud man is always looking down on things and people: and, of course, as long as you are looking down, you cannot see something that is above you.
>
> C. S. Lewis, *Mere Christianity*

Believing that one currently has the best possible understanding of God and science. People are fond of acting as though their specialized expertise somehow privileges them with infallible insights, not only in their own area but also into other areas with which they have no actual experience. Yet knowledge of one person, subject or place is never perfect, and it certainly does not automatically confer intimate knowledge of a different person, subject or place. Nevertheless, it is not unusual to hear scientists

speaking with the presumption of authority about things of
which they know little (all the while criticizing theologians and
philosophers who they imagine are doing the same thing). Consequently, although every thinking person recognizes his or her
susceptibility to blunders of rationality, these become increasingly hard to see in those areas about which one feels most con-
fident. Because Christian scientists inhabit both scientific and
theological worlds where opinions run strong, they must be particularly mindful of the perils in each.

The air of finality that accompanies so much religious pos-
turing, for example, is based on the faulty (but prevalent) notion
that everything of importance one needs to know is already
known. Of all the criticisms leveled by the so-called New
Atheists[3] against traditional religion, narrow-mindedness
(potentially leading to unexamined dogmatism) is perhaps the
most valid. This is something of a paradox, for one would think
that religious people who believe in a transcendent deity would
have the most expansive outlook of all. But strangely the notion
of an infinitely capable God, rather than exposing one's rational
frailties, often manifests itself in a parochial notion of infallible
personal discernment.[4] Somehow the facts that God's chosen
people (Israel in the Old Testament) were regularly mistaken
(e.g., "I desire mercy, not sacrifice," Hos 6:6) and that Jesus' own
disciples frequently got it wrong ("Are you still so dull?" Mt 15:16)
seem to be lost on each new generation, which acts as though it

---

[3]A term used to describe several recent authors who have made especially intense attacks in the name of science.

[4]See Christian Smith, *The Bible Made Impossible* (Grand Rapids: Brazos,
2011).

has finally reached an indisputable answer.[5] Clearly, there is no need to move forward for someone who has already arrived.

This is just as true in scientific arenas, where scientists (Christian or not) are susceptible to elevating their current views to the level of absolute truth, forgetting the tentative nature of any scientific theory, no matter how well it currently seems to be attested. Arthur C. Clarke captured this idea in the first of his three laws: "When a distinguished but elderly scientist says that something is possible, he is almost certainly right. When he states that something is impossible, he is very probably wrong."[6] One of the most interesting and illuminating examples of this is Einstein's resistance to quantum randomness, a stance that this former paradigm breaker maintained until his dying day in the face of the newer quantum paradigm.[7] As noted earlier, Thomas Kuhn has famously suggested that all normal science takes place within some existing paradigm that can serve to further the discipline involved but can also act as a blinder to other possibilities.[8]

---

[5]Historical changes in biblical interpretation (see Peter Harrison, *The Bible, Protestantism, and the Rise of Natural Science* [Cambridge: Cambridge University Press, 2001]) should serve as fair warning that any belief about ultimate theological insight might be regarded as premature.

[6]Arthur C. Clarke, *Profiles of the Future: An Inquiry into the Limits of the Possible* (New York: Harper & Row, 1962), 14.

[7]As Isaacson notes, "In the world of physics . . . Einstein and his fitful quest for a unified theory were beginning to be seen as quaint." Isaacson, *Einstein*, 341.

[8]Thomas S. Kuhn, "The Essential Tension," in *The Third University of Utah Research Conference on the Identification of Scientific Talent*, ed. C. W. Taylor (Salt Lake City: University of Utah Press, 1959); Thomas S. Kuhn, *The Structure of Scientific Revolutions* (Chicago: University of Chicago Press, 1996).

All of this can be particularly difficult for the Christian scientist, who in order to reconcile his scientific and religious beliefs must become a bit of a theologian and philosopher, and therefore must fight the pride war on multiple fronts. However, although rational approaches to religious questions may be honed by scientific training, scientific knowledge itself provides no magic entrance into the club of perfect theological insight.[9]

PROBLEMS WITH A LACK OF INTELLECTUAL HUMILITY

Confidence that one has the best possible view of a subject, scientific or religious, can quickly turn into a tendency to defend that view to the hilt.[10] But such confidence always rests on a set of beliefs, the quality of which can vary extensively. The most dangerous of those beliefs is simply that one is somehow privileged to have obtained a flawless grasp of an issue. The defensive posture that often results—the propensity to protect one's turf at all costs—is one of the first signs of an absence of intellectual humility. Unfortunately, what is so easy to see in others is frequently ignored or suppressed when we look at ourselves.

From such a position it is all but impossible to recognize that the thing being protected may have nothing approaching the truth content or value that has been ascribed to it. Thus today's defense of some imagined ideological sanctuary can in retrospect look more like an attempt to safeguard a dump from benevolent development. The frequency with which this has occurred

---

[9]Consider the apostle Paul's observations in 1 Corinthians 1:18-25.

[10]This tendency is frequently magnified into a full-blown offensive assault on the opposition.

historically might encourage thoughtful scientists to consider more carefully just what it is they are trying to protect.[11]

Nevertheless, the tenacity with which people defend their (possibly tenuous) positions is constantly before us—after all, we are frequently those people. When such stances involve matters at the interface of science and Christianity, confrontations can become noticeably heated, especially when the battle is between what are perceived to be competing sources of knowledge, insight, meaning and truth. If Christians seem more prone to be defensive it is not necessarily because they are less astute than others but because they believe they have more to lose.[12] They are mindful of the admonition in 2 John 8: "Watch out that you do not lose what we have worked for, but that you may be rewarded fully." Certainly, then, some turf deserves protection, but if truth itself—the one sure thing deserving defense—is always something that we can only infer, might not a protectionist stance at least be suspect?

Unfortunately, in a world full of people who want to be right, the most readily deployable and least expensive form of defense for a current set of beliefs is simply an obstinacy to consider the possibility of error. Indeed, the pretense of confidence is sometimes used to cover the fact that there is little if any real foundation for those beliefs. Too many prefer being thought right over being right. It is not, therefore, very difficult to end up

---

[11]Clearly this is a problem that extends beyond science and religion issues. Oppressive opinions, rotten relationships and menacing monarchies have all been objects of the protectionist stance as practiced by individuals, groups and even nations.

[12]See Adam Laats, "To Teach Evolution, You Have to Understand Creationists," *Chronicle of Higher Education*, November 19, 2012.

defending a poor choice of battlegrounds, and even intellectually gifted people can be prone to take a defensive stand to justify an adopted position, sometimes at the expense of poor logic. Examples are not hard to find. The Pharisees' rejection of Jesus, Lamarckian notions of inheritance and Stephen J. Gould's claim that there is no way to reconcile science and religion (and therefore no reason to do so) short of relegating them to separate domains[13] are pertinent illustrations of ill-fated defensive postures (in theology, science and at their boundary, respectively) brought on by a failure to appreciate the alternatives.

If philosopher of science Paul Feyerabend is right, "prejudices are found by contrast, not by analysis."[14] But considering alternatives is precisely what the set mind is unwilling to do. This is just the problem Jesus faced when trying to introduce his audience to new and difficult ideas: "If you were blind, you would not be guilty of sin; but now that you claim you can see, your guilt remains" (Jn 9:41).

> These men are forced into their strange fancies by attempting to measure the whole universe by means of their tiny scale.
>
> Galileo Galilei, *Discoveries and Opinions of Galileo*

Being blind to what is around us both scientifically and theologically is a natural state of affairs, but the results of a failure or unwillingness to recognize that blindness are stagnation and an

---

[13]Stephen J. Gould, "Nonoverlapping Magisteria," *Natural History* 106 (1997).

[14]Paul Feyerabend, *Against Method* (New York: Verso, 1993), 22.

inability to communicate with those who are not so confined. There is always a bit of paradox here because those with diametrically opposed views will always accuse the other of being the one who is blinded. Sometimes they are both right about that.

Intellectual humility, however, also has a sinister side. How does one avoid (to use the old saw) being "humble and proud of it"? As scientists many of us pride ourselves on our open-mindedness, but it is possible to be so open-minded that one is closed to the idea that there are some things about which there may be very little wiggle room. For example, how much can one treat as relative and still maintain a Christian persona? Might there actually be some absolutes that should never be sacrificed or compromised? Most Christians will surely answer this latter question in the affirmative, but knowing what all those areas are can require significant discernment. Indeed, it can take a great deal of humility to acknowledge that a particular interpretive stance might be in need of some revision, but the scientist, of all persons, should at least understand how this works.

ADVANTAGES OF INTELLECTUAL HUMILITY

As mentioned above, Christian scientists are under compulsion to be intellectually humble, not only about both their science and their theology but also about the conjunction of those areas. Let's imagine for a moment that someone—we'll call her Emma— manages to obtain the last word about God. What if Emma is right and everyone else is wrong? Well, hooray for her! This is the stuff of book and movie plots, and while she awaits the endorsements Emma can congratulate herself while reveling in her superiority and contemplating how everyone else could be so blind. Unfortunately for her, there is the small matter of how she is going to convince others that she is the enlightened one.[15] Perhaps she can persuade some by her intellectual prowess or charisma (assuming she has either), but no one—atheist or theist—will be meaningfully changed simply by her assurances that she has the truth. In fact, we have ignored the crucial question of how Emma even knows she has the ultimate answers. Outside our imaginary scenario, it seems far more likely that Emma has but a piece of the truth about God, and probably an extraordinarily small piece at that. Humility is not a natural attribute of atheists or theists, but without it neither will see much reason to search and any truth that might be found will remain beyond their grasp if for no other reason than they are not reaching for it.[16]

In his fascinating account of increases in cosmological knowledge, Richard Panek describes the moment in time when it finally became apparent that "for most of the history of the telescope, astronomers had been studying 1 percent of 2 percent of, at most, 10 percent of what's out there, and calling *that* the

---

[15]As Barbour notes, "There is no uninterpreted revelation." Ian G. Barbour, *Religion and Science: Historical and Contemporary Issues* (New York: HarperOne, 1997), 135.

[16]It is interesting to contemplate what would happen if more atheists and theists were to actively and honestly engage in exploration of their beliefs. One might guess that there would be an influx of people into one camp or the other, but that is not a given, and there may be reasons for suspecting that on average the numbers would remain the same. Unfortunately this is an impossible experiment to run, and not just because of logistical nightmares. There is no way to quantify the suggested "active" and "honest" requirements for such a search or to regulate the experiences of participants or the evidence that is available.

universe."[17] Surely the scientist who thinks deeply about his or her chosen field would echo Panek's sentiments. In areas as diverse as epigenetics, mind-brain-consciousness, evolutionary learning, artificial intelligence and many others we seem to be only scratching the surface of plausible knowledge. But if God has the infinite attributes traditionally ascribed to him by Christians, a comparable shallowness must be true of our theological understanding. As the apostle Paul put it, "Oh, the depth of the riches of the wisdom and knowledge of God! How unsearchable his judgments, and his paths beyond tracing out! 'Who has known the mind of the Lord? Or who has been his counselor?'" (Rom 11:33-34). To acknowledge that one might be wrong, and to admit it when one is wrong, is the gateway to greater discovery. Thus the route to deeper insight—be it scientific, theological or the intersection of the two—begins with intellectual humility.[18]

---

[17]Richard Panek, *Seeing and Believing: How the Telescope Opened Our Eyes and Minds to the Heavens* (New York: Penguin, 1998), 173.

[18]Perhaps one reason Jesus stressed the importance of humility (Mt 23:12; Lk 14:11) was because it enables so many of the other virtues.

PART THREE

SCIENCE AND CHRISTIAN FAITH

7

SCIENCE AND SCRIPTURE

*The intention of the Holy Ghost is to teach us how
one goes to heaven, not how heaven goes.*

GALILEO GALILEI, "LETTER TO THE
GRAND DUCHESS CHRISTINA
OF TUSCANY"

■ ■ ■

MANY CHRISTIANS WORRY that modern scientific theories and Scripture conflict. As we argued in the first chapter, the two books metaphor implies that apparent inconsistencies are the result of human misinterpretation rather than a fundamental disagreement. While this starting assumption helps, it does not resolve inconsistencies when they occur. In this chapter we cannot possibly address all the particular passages that raise questions pertaining to science. We will leave questions about how best to interpret individual verses to biblical scholars, and will instead provide general principles that can be applied to any passage.

PRINCIPLE 1: HAVING THE HOLY SPIRIT AS OUR TEACHER DOES NOT MAKE US INFALLIBLE

To state the obvious: Scripture plays an important theological role in the Christian life. It not only is the primary source of knowledge about the nature of God, but it shows us how to live as Christians. Second Timothy 3:16, for example, explicitly links the inspiration of Scripture with its ability to instruct and correct in righteousness. Moreover, the *doctrine* of Scripture plays an important role as well. For many evangelicals a strong doctrine of Scripture is the best safeguard against future generations slipping into unorthodoxy and even atheism. There is little room for compromise, therefore, if science is seen as undermining the Bible's trustworthiness and authority. But faithful Christian interpreters must also remember: an uncompromising commitment to the inspiration and authority of Scripture does not mean we should have an uncompromising commitment to *our own* interpretation of Scripture. Because we are sinners with imperfect knowledge and motives, we must always be open to the possibility that we have interpreted a verse or passage incorrectly.

Sometimes a lack of openness to other opinions is not rooted in pride but in the belief that the Holy Spirit will guide us to the right opinion, based perhaps on passages like John 16:13, where it is promised that the Spirit "will guide you into all the truth." A preacher on the radio once explained it this way: "Since the Holy Spirit is the author of Scripture, isn't he the best teacher to explain what it means?" The message stressed that we could be most certain in our knowledge when we turn away from human opinions and obtain our teaching from God alone. This idea of the Holy Spirit as the perfect teacher lies behind the quarrelsomeness of the fundamentalist; if for some reason others have trouble perceiving the message of Scripture, one should doubt whether they are truly saved since it is only veiled to those who are perishing. If one's spiritual eyes have been opened to see Scripture as authoritative, then its meaning will not remain obscure.

> Nobody is an infallible interpreter, and we must always stand ready to reconsider our interpretations in light of new information. We must not let our interpretations stand in the place of Scripture's authority and thus risk misrepresenting God's revelation. We are willing to bind reason if our faith calls for belief where reason fails. But we are also people who in faith seek learning. What we learn may cause us to reconsider interpretations of Scripture, but need never cause us to question the intrinsic authority or nature of Scripture.
>
> John Walton, *The Lost World of Genesis One*

But this way of framing biblical interpretation as picking either the Holy Spirit or human opinion can lead to an overconfidence in one's own opinions. For example, a student of Martin Luther reported that he said this about Copernicus's theory that the earth orbits the sun: "So it goes now. . . . Whoever wants to be clever . . . must do something of his own. This is what that fellow does who wishes to turn the whole of astronomy upside down. . . . I believe the Holy Scriptures, for Joshua commanded the Sun to stand still, not the Earth."[1] Luther frames the problem as whether one should believe human opinion or divine revelation. Put so starkly, why would any Christian not pick divine revelation? In retrospect, however, we can see that Luther's way of putting the issue made it difficult for him to give Copernicus a fair hearing. The larger point is that even though Christians have the Spirit of God, we still can err when reading the Bible.

PRINCIPLE 2: WE MUST READ THE BIBLE IN COMMUNITY

The second principle builds on the first. Our own fallibility means we should be open to the perspectives of other Christians, who also have the Holy Spirit as a teacher.[2] The assumption of some Christians is that human opinion can only interfere with a true understanding of Scripture. The solution to every theological controversy is thus to cast away human opinions and let the Bible speak for itself, letting there be "no creed but the Bible." The prominence of the "no creed but the Bible" tradition in American evangelicalism is motivated by the desire to remove all human influence from our interpretation of the Bible, letting the Holy Spirit alone speak to us through the text. To admit the role of creeds in the Christian faith is therefore believed to imply that the Bible is in some way unclear and to mistakenly assume that the biblical message needs to be restated and clarified so that others can grasp it.

---

[1]Owen Gingerich, *The Book Nobody Read: Chasing the Revolutions of Nicolaus Copernicus* (New York: Bloomsbury, 2009), 136.

[2]Josh Reeves, "Theology and the Problem of Expertise," *Theology Today* 69, no. 1 (2012): 39.

This individualistic way of approaching Scripture influences the way some people think we should obtain our beliefs more generally. Institutions such as the church or the university, we are often told, are more interested in accumulating and holding power than finding truth. A frequent motif in Western culture is that the best way to find what is true is to "forget tradition, ignore authority, be skeptical of what others say, and wander the fields alone."[3]

Institutions like the church or university can often be corrupt, of course, but individualists fail to appreciate an equally important point: our minds are often too weak to find truth for ourselves. We accept beliefs for bad reasons; we accept too many answers that fit with our own biases; we accept easy answers when we should keep searching. Individualist epistemologies mistakenly assume a viewpoint of epistemic egalitarianism— that all members of the community are equally competent and that there are no significant limits on each one's ability to investigate questions.[4] Yet attention to real communities reveals that this is hardly the case—for a vast majority of church history Christians have lacked basic literacy, much less extensive knowledge of Scripture. It has even become something of a trend in evangelical literature to lament the lack of theological knowledge in evangelical communities. The source of the lament is regularly born out in surveys, such as a recent Pew poll showing that three out of ten evangelicals could not name all four

---

[3]Steven Shapin, *The Scientific Revolution* (Chicago: University of Chicago Press, 1996), 69.

[4]Alvin I. Goldman, "Experts: Which Ones Should You Trust?," *Philosophy and Phenomenological Research* 63, no. 1 (2001): 85.

Gospels.[5] From a theological point of view, the same sin that infects and corrupts institutions also infects and corrupts individual hearts and minds.

> The debate that has been conducted in terms of "creation versus evolution" has gotten caught up with all kinds of other debates (in American culture in particular), and this has provided a singularly unhelpful backdrop to the would-be serious discussion of other parts of the Bible.
>
> N. T. Wright, *Simply Christian*

We should have little hope of interpreting the Bible well without the assistance of others, just as there is little hope of becoming a scientist on one's own. The best way to think about our relationship to Scripture is in terms of discipleship, where one's ability to read the Bible is slowly transformed under the guidance of others, just as Jesus gathered around him a community of followers in order to lead them to a fuller understanding of the truth. The analogy of discipleship suggests that reading Scripture is a difficult thing to do well and cannot be accomplished without gaining virtues such as humility, truthfulness and charity, which only come from interacting with those around us. As one seeks to puzzle out ways of reconciling science and Scripture, it is of the utmost importance to find quality teachers, those who combine intellectual rigor with the virtues that come from a life of Christian faith.

---

[5]Pew Forum on Religion and Public Life, *U.S. Religious Knowledge Survey* (Washington, DC: Pew Research Center, 2010), 21.

PRINCIPLE 3: NOT JUST
A LITERAL INTERPRETATION

Sometimes Christians suggest we can avoid interpretative errors if we simply adopt a literal reading of the Bible, immortalized in the bumper sticker "God says it, I believe it, that settles it." Yet even if we ignored all the genres of the Bible that are inherently difficult to interpret (e.g., poetry and prophecy), the meaning we get out of the text is shaped by our background assumptions that we bring. In a famous metaphor of twentieth-century philosophy, gaining knowledge is like modifying a boat as we are sailing at sea; we can tinker with any piece of the boat we wish, but we cannot replace the whole without sinking. The metaphor's lesson is that our background assumptions provide the structure that makes reasoning possible. Thus when considering a passage of Scripture, we cannot separate our cultural and theological assumptions from the interpretation we make.

The same sort of simplistic interpretative scheme has sometimes been advocated in science. Some have argued that the job of the scientist consists mainly of collecting and arranging "facts"—nuggets of truth that are uncontaminated by our personal beliefs. As we discussed in a previous chapter, no philosopher of science seriously holds to this after the work of Thomas Kuhn, who showed how particular interpretations depend on larger paradigms to make sense. This does not mean that anything goes in what scientists claim about nature, but it does mean that scientific theories are hardly a "literal" reading of nature.

The recognition that interpretation is always an interplay between the text and our assumptions should drive us to interpret the Bible with humility and charity and motivate us to consider other positions with openness. Failure to read the Bible with humility might not only lead one to make interpretative errors, but it can also undercut the witness of the larger church. The explosion of denominations since the Protestant Reformation—where every Christian was encouraged to read the Bible for themselves—has often been a consequence of an inflexible approach to biblical interpretation, failing to acknowledge that equally faithful Christians might disagree. The ability to read Scripture for ourselves is undoubtedly a good thing, but the quarrelsomeness and factions that come with Bible reading have been a negative for the witness of the church.

> The Holy Spirit had no intention to teach astronomy; and in proposing instruction meant to be common to the simplest and most uneducated persons, he made use by Moses and the other prophets of popular language that none might shelter himself under the pretext of obscurity.
>
> John Calvin, *Commentary on Psalms*

An inflexible approach can also create unnecessary roadblocks for those considering the Christian faith. For example, insisting that only one interpretation of the opening chapters of Genesis is authentically Christian can push outsiders away—an unfortunate result given that Christians have always interpreted those chapters in multiple ways. As Christians in the first few centuries acknowledged, the variety of plausible ways of interpreting Genesis is attributable as much to the poetic nature of the text itself as to human sinfulness. Even Augustine argued over 1,500 years ago in his commentary on Genesis that an overly literal approach leads to interpretative problems. How, he asks, can the "days" of Genesis be solar days if the sun was not created until day four?[6]

PRINCIPLE 4: TO KNOW WHAT THE BIBLE MEANS FOR US TODAY, WE SHOULD FIRST UNDERSTAND WHAT THE BIBLE MEANT TO ITS ORIGINAL AUDIENCE

The most likely way to err with respect to biblical interpretation is to fail to interpret the Bible in its cultural context. In other words, we fail to recognize what the passage would have meant to those who first heard the message. Of course, sometimes a passage seems so straightforward that a consideration of context seems hardly necessary. The commandments "Do not steal" or "Do not murder," for example, seem to have a clear meaning across cultures, though it may be helpful to know how the principles were applied in ancient Israel and the early church. When we are trying to discern the meaning of a difficult or controversial biblical passage, however, the most important step is to consider the passage in its cultural context. As the Old Testament scholar John Walton says, "God's Word was written for us, but not to us. Bringing the ancient text to modern readers is not just a matter of word rendering; it's also a matter of understanding the culture in which the text was written."[7]

---

[6] Augustine, *The Literal Meaning of Genesis*, bk. 1.

[7] John H. Walton, *The Lost World of Genesis One: Ancient Cosmology and the Origins Debate* (Downers Grove, IL: InterVarsity Press, 2010), 9;

The reason modern readers have to understand the context of a biblical passage is that God's revelation is accommodated to the understanding of those who first heard it. The principle of accommodation says that God communicates revelation in terms that the audience of the day will understand. Theologians throughout church history, including Augustine, Aquinas and Calvin, have affirmed this principle. As Calvin said, Scripture "proceeds at the pace of a mother stooping to her child, so to speak, so as not to leave us behind in our weakness."[8] Some truths would be too overwhelming or complex for the ancient Israelites or the first-century followers of Jesus to understand.

> But because God chose to speak his work through human words in history, every book in the Bible also has historical particularity; each document is conditioned by the language, time, and culture in which it was originally written (and in some cases also by the oral history it had before it was written down).
>
> Gordon Fee and Douglas Stuart,
> *How to Read the Bible for All Its Worth*

While the principle of accommodation is deeply ingrained in Christian theology, existing well over a thousand years before the Scientific Revolution, it has important implications

---

Kevin P. Emmert, "The Lost World of Adam and Eve," *Christianity Today* 59, no. 2 (2012): 42.

[8]John Calvin, *Institutes of the Christian Religion* (Peabody, MA: Hendrickson, 2007), 3.21.4.

for reading Scripture in light of modern science.[9] One implication is that God did not give to ancient cultures a scientific understanding beyond the cultures around them. In other words, if the surrounding ancient Near Eastern cultures believed the Earth sat on pillars, then the Israelites did too (e.g., Job 9:6). If the surrounding cultures believed that the heart was the organ for thinking, then the Israelites did too (e.g., Gen 24:45).

Does this make the Bible untrue? Definitely not! Consider the opinion of Charles Hodge, a leading nineteenth-century theologian who is known for his defense of the Bible being "without error." He said,

> As to all matters of science, philosophy, and history, [the sacred writers] stood on the same level with their contemporaries. They were infallible only as teachers, and when acting as the spokesmen of God. Their inspiration no more made them astronomers than it made them agriculturists. . . . We must distinguish between what the sacred writers themselves thought or believed, and what they teach.[10]

Hodge himself argued, for example, that the writers of Scripture believed the sun moved around the earth but they nowhere taught this as part of Christian doctrine. To insist that the Bible writers had perfect knowledge of science is to assume that science gives us the most superior type of knowledge. By contrast, the purpose of the Bible was to convey spiritual knowledge,

---

[9] Alister E. McGrath, *Christian Theology: An Introduction* (West Sussex, UK: Wiley-Blackwell, 2011), 192.

[10] Charles Hodge, *Systematic Theology: Volume One* (New York: Charles Scribner, 1871), 165, 171.

especially the character of God as revealed in Christ Jesus, on which the Bible is the supreme authority.

There is so much more that can be said about biblical interpretation, of course. But as a general approach we hope these principles will orient you when dealing with difficult passages.

8

ARE SCIENTISTS
MOSTLY ATHEISTS?

The fool says in his heart, "There is no God."

PSALM 14:1

■ ■ ■

THE ATHEIST, OF COURSE, thinks Psalm 14 is quite back-wards: "The fool says in his heart, 'There is a God.'" Quite frankly, Christians are warned of the dangers of calling anyone a fool (Mt 5:22), and in any event it is not a helpful way to approach the differences in responses to the question of God's existence.

That those differences exist is obvious. Survey results in-volving members of the American Association for the Ad-vancement of Science, for example, reflect them and also show that the religious beliefs of that group of scientists do not mirror those of the population at large.[1] The perception of the atheistic tendencies of scientists is further buttressed by public stances

---

[1]See David Masci, "Scientists and Belief," Pew Research Center, Novem-ber 5, 2009, www.pewforum.org/2009/11/05/scientists-and-belief. Of course one might also wonder whether the preponderance of atheists in the AAAS merely means that atheists are more likely to join.

periodically taken by well-known scientists who are also atheists. In a recent book, *Science vs. Religion: What Scientists Really Think*, sociologist Elaine Howard Ecklund surveyed almost 1,700 scientists (including social scientists) at elite research universities. She found 50 percent of scientists in her sample to be members of a religious tradition, even though 34 percent were atheists and 30 percent were agnostics.[2] Despite the diversity of scientists' beliefs, her interviews and surveys revealed a strong social pressure to keep religious views private.[3] In any case, among those scientists who are not atheistic, only some are Christian. Of course one might also conduct a survey to determine whether scientists are primarily male or female, or belong mainly to one ethnic group or another, but such questions do not seem nearly as significant as asking about the role of religion in their lives. There are several reasons for thinking this is the case.

In the first place, any scientist is going to be at least somewhat attuned to the prevailing climate of his or her discipline. As described earlier, no scientist can work in a true vacuum, and knowing that one's general opinions about any perceived reality are shared by peers provides a level of psychological comfort that extends far beyond the sciences. For the scientist, however, the differences between one's own religious beliefs and those of many in the discipline can be a source of major concern because the

---

[2]Elaine Howard Ecklund, *Science vs. Religion: What Scientists Really Think* (New York: Oxford University Press, 2012), 16. The disparity in numbers is attributable to scientists who identify culturally with a religious tradition without believing its doctrines.
[3]Ibid., 24.

presumed rationality represented in the sciences is often pitted against religious belief, and the conclusions derived from the sciences are frequently paraded as a substitute for any alternative forms of knowledge.[4] Religiously inclined scientists can therefore be left wondering whether they might be mistaken about their religious beliefs or whether they are in greater danger of losing their religious faith than they would be in a different profession.

Now, asking whether one's faith in something rests on a solid foundation is a good question for anyone, theist or atheist.[5] For the Christian scientist who perceives herself in the minority, however, it may also be helpful to remember that although scientific consensus is usually deemed the hallmark of a good theory, actual progress in science has frequently involved rejecting the majority view. When it comes to worldviews that potentially transcend science, this observation becomes especially significant. In other words, it does not logically follow that because individuals have mastered one limited domain of human understanding they are expert or even competent in other domains. A person may understand the concept or even the value of physical fitness, for example, without actually being physically fit. One might suppose the same applies to spiritual fitness, and that individuals who have developed significantly in their scientific understanding might yet be quite naive theologically. It is thus unnecessary to conclude that a scientist cannot be a Christian, or vice versa.

---

[4]See J. Wentzel Van Huyssteen, *The Shaping of Rationality: Toward Interdisciplinarity in Theology and Science* (Grand Rapids: Eerdmans, 1999).
[5]See Steve Donaldson, *Dimensions of Faith: Understanding Faith Through the Lens of Science and Religion* (Eugene, OR: Cascade, 2015).

> A discovery of the divine does not come through experiments and equations, but through an understanding of the structures they unveil and map.
>
> Antony Flew, *There Is a God*

Nevertheless, while the Christian scientist may find special reason to believe in God because of the view provided by science, agreeing with the psalmist that we are "fearfully and wonderfully made" (Ps 139:14), the fact remains that many scientists fail to see God in the natural order. It is important to try to understand this and to explain other factors that prevent scientists from being theists. Ultimately, however, the main reasons a scientist might be an atheist come down to too large a view of science, too tired a view of religion and too lofty a view of humans (and their success in science). These views are what lead to proclamations such as that by cosmologist Lawrence Krauss that science is "an atheistic discipline."[6] We'll consider each view in turn.

TOO LARGE A VIEW OF SCIENCE

A common misconception is that whereas God was once needed to explain the operation of the universe, science has relentlessly taken over that task with the result that there is no longer any need to invoke deity.[7] What non-scientists and scientists alike

---

[6]Lawrence Krauss, "God and Science Don't Mix," *Wall Street Journal*, June 26, 2009.

[7]This is the God-of-the-gaps view: "There are reverent minds who

often fail to appreciate is that rather than destroying the mystery, science simply changed its locus. While it is clear that science has been eminently successful in establishing predictive and explanatory frameworks for the natural order, it is by no means clear that the mystery is gone. As Einstein famously noted, "The eternal mystery of the world is its comprehensibility."[8] Failure to appreciate this has rendered science the new god for many. Coupled with the fact that nature is the old god for many others, one is left wondering whether there is in fact any such thing as a true atheist.

> Perhaps the desire to make God into a domestic craftsman is because he is more easily tamed that way.
>
> John Polkinghorne, *The Way the World Is*

In any case, rejecting God because of the success of science is the result of assuming that one of the primary purposes of science is to substitute for God. Yet there is nothing in either science or theology to suggest that should be the case. Nevertheless, when anything is unduly exalted, it can be made to serve any purpose one wishes. This was certainly the case with ancient deities and is just as true of science when given godlike status.[9]

---

ceaselessly scan the fields of Nature and the books of Science in search of gaps—gaps which they will fill up with God. As if God lived in gaps?" Henry Drummond, *The Lowell Lectures on the Ascent of Man* (Radford, VA: Wilder, 2008), 171.

[8]Albert Einstein, "Physics and Reality," trans. Jean Piccard, *Journal of the Franklin Institute* 221, no. 3 (1936): 351.

[9]Donaldson, *Dimensions of Faith*, 192-200.

In the end, of course, one will serve whatever it is that has become his god.[10]

There is quite a paradox here because, despite its many merits, the science that would be god is in fact limited in a variety of important ways. First, we must acknowledge that science is a human endeavor and is hence constrained by its practitioners, who come to their tasks with imperfect cognitive capacities, relatively short life spans and susceptibility to logical missteps.[11] Add to those restrictions the facts that observation, measurement and theory formation are never exact and always occur in a particular context, and that no one is capable of mastering and maintaining a comprehensive view of even a small piece of scientific knowledge, and it is no surprise that scientific theories have been a moving target.[12]

In addition, a number of individuals, irrespective of their religious inclinations or lack thereof, have begun to call attention to the problems of placing an undue focus on the purely reductionist aspects of science as conventionally conceived.[13] Despite

---

[10]As the author of 2 Peter put it, "People are slaves to whatever has mastered them" (2 Peter 2:19).

[11]Donaldson, *Dimensions of Faith*, 79-90.

[12]Ibid., 57-63, 90-112. C. S. Lewis suggests that "when changes in the human mind produce a sufficient disrelish of the old Model and a sufficient hankering for some new one, phenomena to support that new one will obediently turn up." *The Discarded Image: An Introduction to Medieval and Renaissance Literature* (Cambridge: Cambridge University Press, 1964), 221. Although it is probably not quite so simple as that, Lewis's observation has much to commend it.

[13]See Thomas Nagel, *Mind and Cosmos: Why the Materialist Neo-Darwinian Conception of Nature Is Almost Certainly False* (New York: Oxford University Press, 2012); Terrence Deacon, *Incomplete Nature: How Mind Emerged from Matter* (New York: W. W. Norton, 2013);

these and other limitations, science has been successful enough to win the worshipful adoration of many. Sometimes this is not because science is viewed as exceptionally special but because religion is viewed as especially lacking.

TOO TIRED A VIEW OF RELIGION

It seems strange, but scientists who understand quite well the dynamic nature of the scientific process are frequently reluctant to grant the same grace to theological understanding. In other words, while our scientific conceptions of the natural world are allowed to grow and such growth is actively promoted, changing views of God, divine action or other "ultimate realities" are deemed a sign of weakness in religion in general and whatever religion is under consideration in particular.

Nothing in the nature of religious or general understanding, however, warrants such a position. Certainly the Judeo-Christian tradition reflects a growing understanding of God's attributes and his expectations for human conduct, much of which occurred long before the rise of modern science. Nevertheless, Christian and non-Christian alike sometimes fail to acknowledge the interpretive possibilities inherent in (for example) biblical exegesis, a state of affairs that may simply hinder Christian development but can also shut the door to atheists who might otherwise have made it into the kingdom by a more circuitous than normal route. It seems strange, for instance, that

Robert Ulanowicz, *A Third Window: Natural Life Beyond Newton and Darwin* (West Conshohocken, PA: Templeton Foundation Press, 2009); Sandra Mitchell, *Unsimple Truths: Science, Complexity, and Policy* (Chicago: University of Chicago Press, 2012).

someone like Darwin who was adept at perceiving new ways to interpret the biological record seems to have been unable to make comparable concessions to the interpretation of Scripture or his own preconceptions.[14]

The ultimate problem here eventually manifests itself as too small a view of God. Consider, for instance, the conveniently small god proposed by Neil deGrasse Tyson: "If I propose a God . . . who graces our valley of collective ignorance, the day will come when our sphere of knowledge will have grown so large that I will have no need of that hypothesis."[15] Now, it is difficult to have a smaller view of God than none—the atheist view—but that view itself arises in large part either because it has become impossible for atheists to imagine how any god could answer the deep questions they have posed for him or because no such questions are being asked. Yet attempting to shrink God is perhaps not so strange for someone who overly exalts a reductionist approach to understanding.

TOO LOFTY A VIEW OF HUMANS

If God is made small enough, any human endeavors (and consequently the humans who undertake them) begin to look overly significant. Thus an undue pride in human knowledge

---

[14]See Karl Giberson, *Saving Darwin: How to Be a Christian and Believe in Evolution* (New York: HarperOne, 2008), regarding the impact the traditional doctrine of hell as well as his personal struggles with theodicy had on Darwin's theological convictions.

[15]Neil deGrasse Tyson, "Holy Wars: An Astrophysicist Ponders the God Question," in *Science and Religion: Are They Compatible?*, ed. Paul Kurtz (New York: Prometheus, 2003), 79. This should sound familiar to anyone who has heard Laplace's famous rejoinder to Napoleon.

and accomplishments can eclipse God, rendering him all but invisible, as a finger held too closely to one's eye can block the sun. We've already discussed some of the problems associated with a lack of intellectual humility, but it is difficult to imagine greater problems than those arising within an individual who imagines that she can discover God on her own terms, or who believes that his special gifts in scientific understanding somehow privilege him above others in obtaining a relationship with God—or that those insights simply remove the need for God altogether.

> Humans persistently abandon their capacity for finding truth in favor of abuses that spring from idolatrous self-interest.
>
> Mark Noll, *Jesus Christ and the Life of the Mind*

It is probably also a mistake to think that just because someone desires truth in one arena they are anxious or willing to pursue it in others.[16] Attempting to discover scientific truth is perceived as a discovery process that can undergird a career and lead to prestige as one masters a discipline, but trying to discover the truth about God can lead to worries about the need to sacrifice other things held dear, fear that one will be rejected by peers who don't share the same religious views, or concerns that one might actually find good reasons to believe in a God whose nonexistence is assumed to be more convenient. Courageous scientists are not necessarily courageous explorers in other

---

[16]Donaldson, *Dimensions of Faith*, 245-50.

domains. How many atheists, for example, are asking, If there is a God, do I want to know?[17]

A Case Study

To illustrate the interplay of these three roadblocks to theism, consider the following famous quotation from the eminent evolutionary biologist J. B. S. Haldane:

> My practice as a scientist is Atheistic. That is to say, when I set up an experiment I assume that no god, angel or devil is going to interfere with its course; and this assumption has been justified by such success as I have achieved in my professional career. I should therefore be intellectually dishonest if I were not also Atheistic in theory, at least to the extent of disbelieving in supernatural interferences in the affairs of the world.[18]

Despite the sound-bite appeal of such a proclamation or the scientific credentials of the speaker, this is a textbook example of employing lavishly loose logic to support a preexisting bias. Consider, for instance, the logic of such a statement with a few simple substitutions:

---

[17]The reverse question is only fair: How many Christians are willing to ask themselves, If there is no God, do I want to know? Difficult as these questions can be to ponder, they are critical to helping frame issues of faith and reason. A really interesting question thus arises in light of these considerations: Does being smart and knowledgeable work against knowing or believing in God? It would be easy to simply think that in discussing God and human wisdom in 1 Corinthians 1 Paul answers that question, but that would be to misunderstand Paul and also to ignore what he goes on to say in (for example) 1 Corinthians 2.

[18]J. B. S. Haldane, *Fact and Faith* (London: Watts & Company, 1934), vi-vii.

My practice as a runner is selfish. That is to say, when I run I assume that no one is going to interfere with my exercise and that no one but me is going to reap physical or psychological benefits from it; and this assumption has been justified by such success as I have achieved in my running life. I should therefore be intellectually dishonest if I were not also selfish in theory, at least to the extent of believing selfishness was justified in worldly affairs.

Now, it seems apparent that in virtually all of our activities (intellectual, physical, relational and so forth) we ignore those elements that we believe are extraneous to them, but that does not thereby render those elements meaningless in other contexts. The fact that music plays no role in his diagnosis does not lead a physician to conclude that music lacks value in other domains (although that may individually be true for him). Neither does a swimmer discard the idea that legs are useful for walking just because they have a different use while she is trying to swim the English Channel. In fact, because many of the successes we have had in our lives resulted without any influence from science, by Haldane's logic we would be justified in eschewing science in all areas. To put it another way, if I perceive no power in science to improve my tennis game, engender love for a particular musical genre or enhance interpersonal relationships, then why is it not reasonable for me to assume that it is useless in all other affairs of the world? Clearly all these conclusions are non sequiturs, but when one takes too large a view of science and humans and too small a view of God, conclusions such as Haldane's are not uncommon.

Certainly there seems little danger of God becoming any smaller for an atheist, but invariably something else will then be elevated into an object of devotion. When that turns out to be science, there is the potential for an interesting form of "science of the gaps" where the concern is to defend those areas in which scientific explanation is fundamentally inadequate. The resulting protectionist stance is reminiscent of that accompanying a lack of intellectual humility; indeed, it is an unwillingness to express such humility that can constitute the barrier that prevents the atheist from entering the kingdom.[19]

> The skeptic was quite right to go by facts, only he had not looked at the facts.
>
> G. K. Chesterton, *Orthodoxy*

What Does It Matter?

For a number of years it has been difficult for anyone driving across the Texas panhandle on Interstate 40 to miss seeing what is billed as one of the largest crosses in the Western Hemisphere. The metal structure is visible from miles away across the flat plains, but recently a large number of wind turbines have visually diminished its prominence. The spiritual significance of the cross remains unchanged, however, no matter how many structures of a different nature exist around it.

In an age when numerous individuals (including some prominent scientists) seem to represent something radically different from traditional Christian beliefs, it is easy to forget that many of the early and most famous scientists were devout Christians.

---

[19]See, e.g., Matthew 18:3: "Unless you change and become like little children, you will never enter the kingdom of heaven."

As we've seen, this included such figures as Galileo, Kepler, Newton and Boyle. In fact, there is good reason to think that modern science took root and grew best in a Christian setting, despite having had ample opportunity to do so in other environments.[20] But regardless of historic or current precedent, the Christian scientist is ultimately confronted with Jesus' claim that "wide is the gate and broad is the road that leads to destruction, and many enter through it. But small is the gate and narrow the road that leads to life, and only a few find it" (Mt 7:13-14). What, then, does it matter to the faithful Christian if many or even most of his peers are atheist, or just agnostic? Actually, as we are about to see, it matters a great deal, but not because the Christian should be worried about conforming to a strictly secular standard. The Christian scientist, it turns out, has a special opportunity to bridge the gap between Christian belief and that of a largely non-Christian world.

---

[20]See Alister E. McGrath, *Science and Religion: A New Introduction* (West Sussex, UK: Wiley-Blackwell, 2010); Peter Harrison, *The Bible, Protestantism, and the Rise of Natural Science* (Cambridge: Cambridge University Press, 2001).

9

SCIENCE FOR THE GOOD
OF THE CHURCH

*Continue to work out your salvation with fear and
trembling, for it is God who works in you to will
and to act in order to fulfill his good purpose.*

PHILIPPIANS 2:12-13

■ ■ ■

ALISON GOPNIK, A PSYCHOLOGIST and philosopher at
the University of California, Berkeley, has claimed that "consciousness narrows as a function of age. As we know more, we
see less."[1] A variety of factors might make this true, including
work-related expectations, paradigm blindness, comfort with
existing beliefs, fear of change, basic human limitations, and
pressure from social, religious and professional support groups.
But Gopnik's "we" is not just restricted to individuals, and her
statement can characterize religious communities and whole

---

[1]Alison Gopnik, "Why Babies Are More Conscious Than We Are" (lecture at Toward a Science of Consciousness conference, Tucson, AZ,
April 12, 2008).

societies as well. One result of narrowing consciousness is a tendency to select only evidence that supports what is already believed—one can end up accumulating cherry-picked knowledge that enhances certainty about something even though it is wrong. At one time there may have been a willingness to question something that has now become so entrenched as to be all but unassailable. The result? We see less.

> Crises of faith can engender unquestioning acceptance of the current situation or lead us to abandon a belief altogether but they can also turn us into explorers.
>
> Steve Donaldson, *Dimensions of Faith*

This need not, however, be a foregone conclusion. Although individual frailties can cascade into an entire culture such as the church, so can individual strengths. Thus as a church matures it may either be plagued with tunnel vision or become visionary. The ever-present danger is that today's vision becomes tomorrow's rut. There is no rest for those who would prevent an accumulation of knowledge and insight from becoming a prison. That Jesus' harshest criticisms were directed at the religious elite of his day should be adequate reminder of the vulnerability of any religious person—Christians included—to this threat.

Consequently, every church member is under obligation to understand and guard against a paralysis of perception and thought that can make spiritual progress impossible. What we would like to suggest here is that the Christian who is also a scientist is in a special position to recognize this as a potential problem and also to help address it. For one thing, numerous intellectual issues that face modern Christians arise at boundaries where science and religion meet. Christian scientists will in many cases already have begun to struggle with melding what have often been seen as disparate disciplines, asking, How do we make sense of the various claims of science and the Christian faith? Fortunately the same critical analysis skills that enable a successful scientific career are also useful for helping to build bridges between the two areas.

In attempting to live out Paul's injunction, "Whatever you do, do it all for the glory of God" (1 Cor 10:31), the Christian scientist has a distinctive opportunity to use his or her unique gifts to be salt and light (Mt 5:13-16), both within and beyond a specific Christian community. Those efforts can in turn function as integral components in the overall growth and development of a congregation. It would be a serious mistake to assume that science and Christianity are somehow at odds or that how they interact is irrelevant to the church's real mission and ministry.

VENUES FOR FAITHFUL LIVING

In this section we'll consider a variety of ways in which the Christian scientist can positively affect the mission and ministry of the church.

Getting personal. If anyone, scientist or not, is to have a vibrant influence in their church, it must begin with how they conduct their personal lives. This seems so obvious as to deserve scant attention, but public perceptions about scientists make it an especially relevant issue for those who are Christian. It may or may not be true that most scientists are atheists, but it is especially important that the Christian scientist is not mostly atheistic. It is possible for a scientist to profess a Christian commitment but to live in such a way that no one can tell.

One of the key ways to dispel the illusion that science is an atheistic discipline is to demonstrate that one can lead a faithful Christian life at home and at work—that a certain well-roundedness can be maintained that not only doesn't exclude God from family or career but positions him front and center. The Christian scientist must operate professionally under the assumption that there is an order to natural processes, but unlike Haldane feels no compulsion to act as though that somehow repudiates God.[2] Furthermore, although there will always be things outside the control of any individual, the scientist should model how to react to adversity with appropriate Christian responses no less than any other Christian. These attitudes and behaviors are the starting points for faithful Christian ministry.

Sharing the benefits of science. Despite the problems that have accompanied development of industrial and technological societies, even if it were possible to return to an earlier time, few individuals would be willing to do so—for the simple reason that the benefits made possible through science are deemed to outweigh the perils. Many of those benefits play directly into the ministry of the church, and scientists are in an ideal position to help make that clear (especially to those persons who seem to reject the idea that science can contribute anything positive to church practice).

Jesus, for instance, has been called the Great Physician, but many more people have been cured through the application of

---

[2]In formal terms, methodological naturalism does not entail atheism.

sound medical science than he ever healed in his earthly ministry—something he may have had in mind when he said his disciples would do greater things than he had done (Jn 14:12). As Christians have recognized since the time of Christ (and understand from his example), it is difficult to share a spiritual message with a starving, sick or hurting person. In other words, if something fundamental is absorbing a person's attention, it can block the deeper messages that the church wishes to bring. Yet science has provided a way to address many of the physical, mental and emotional ills that plague individuals and promises to address even more in the years ahead. Thus as technologies enabled by scientific understanding have contributed to increased food production, clean water and even an ability to spread the gospel through a variety of electronic means, science has partnered (perhaps, but not necessarily, unwittingly) with Christianity in some of its most significant ministry objectives.

Helping to educate ministers and congregations. In recounting his conversion experience, noted Watergate conspirator Chuck Colson describes how when he was at one of his lowest moments a friend reached out and took from a shelf a book that he thought might change Colson's life. The book was C. S. Lewis's *Mere Christianity*, and anyone who has read it can immediately understand why it had the impact on Colson that his friend had hoped. The key thing to note here is the perhaps surprising fact that it was not the Bible that was given to Colson. We relate this story not to discount the importance of the Bible— indeed without its message books such as *Mere Christianity* are meaningless—but to point out that God can speak in any number of ways.

Although Christian tradition maintains that the Bible provides the necessary knowledge about God's saving grace, any idea that the Bible is a complete source of all knowledge, even about God, is impossible to justify. A God of infinite attributes cannot be fully described in a finite book, and there is no reason to think that God would limit himself to a single source. If he did, there would seem to be little need for preachers or Bible teachers. But because God can speak in various ways, it is useful to ponder (in the context of this book) how he might do so through Christian scientists.

The fact is, modern science has provided a different view of the world than was available to the people who wrote and originally read the various parts of the Bible (which were different peoples at different times). The more informed and conversant Christians become in areas that are touched by both scientific and theological reflection, the more they can contribute from these particular vantage points to the overall ministry objectives of the church. Clearly not everyone will reach the same level, nor need they, but it would be a shame if all church members—and particularly ministers—did not have a basic understanding about the key issues and not merely some ill-formed opinions.

A description of science cast entirely in terms used by scientists would be incomprehensible to outsiders.

Bruno Latour and Steve Woolgar,
Laboratory Life

Scientists in congregations can help provide insight into relevant areas both formally (via scheduled lectures, discussion groups, book reviews, addendums to Bible studies and the like) and informally as they simply interact one-on-one with members of the congregation and ministerial staff or share their perspectives in existing group settings. A good starting place is to address the fundamental question, Science and religion: are they compatible? (which is the title of at least two books reflecting a deep suspicion among many that they may not be).[3] As described later, additional emphasis can be placed on other key questions at the intersection of science and religion in general and Christianity in particular.

It is disconcerting that there is an apparent disconnect between scientific and theological perceptions among members of many congregations. Furthermore, the general track record of many denominations and sects has not been especially attractive with respect to productively assessing the fruits of modern scientific thought as it pertains to theological perspectives, even though the rewards of doing so are potentially great. Cultivating a richer, deeper engagement between science and Christian thinking can help church members better understand the importance of integrating these areas, both with respect to their own spiritual development and to how they are perceived by those outside the church. In short, the goal is to make the mind a full partner with the heart, soul and strength in loving and serving God.

Reaching the intellectually disenfranchised. There are any number of reasons why people may have chosen not to accept

[3]Daniel C. Dennett and Alvin Plantinga, *Science and Religion: Are They Compatible?* (New York: Oxford University Press, 2011); Paul Kurtz, ed., *Science and Religion: Are They Compatible?* (New York: Prometheus, 2003).

the Christian message or to engage with a local church, but in an increasingly educated society one of those reasons involves the perception that there are intellectual barriers between scientific and religious ways of thinking that are simply insurmountable. A hunger for understanding can create a void that, for some people, has distracting results analogous to those that physical hunger has for others. Being able to show how science and Christianity are compatible is a way to knock down barricades and build bridges with such marginalized people who think science has somehow eliminated the need for God. If the Christian scientist does not attempt to address those concerns, who will?

> Opponents must be gently instructed, in the hope that God will grant them repentance leading them to a knowledge of the truth.
>
> 2 Timothy 2:25

Not surprisingly, the best hope for reaching an atheist or agnostic may be the informed, caring, Christian scientist who is prepared not only for honest and educated discussion but also to pray for those whose ideas currently differ from her own. Many Christians see atheistic scientists such as Daniel Dennett, Richard Dawkins, Sam Harris and E. O. Wilson as the enemy, but even if that is the case, it would be prudent to recall that Christians are under injunction to pray for their enemies (Mt 5:44). In trying to instantiate Paul's vision of becoming "all things to all people" to win some (1 Cor 9:22), the Christian scientist is also living out the call to "always be prepared to give an answer to everyone who asks you to give the reason for the hope that you have" (1 Pet 3:15).[4]

SPECIFIC WAYS THE CHRISTIAN SCIENTIST CAN HELP THE CHURCH

There are a variety of ways the Christian scientist can help church ministers, members and prospective members understand relationships between science and Christianity as they pertain to personal and corporate growth and development.

Clarifying the role of context in the development of religious and scientific faith. From table manners and marriage rituals to scientific and religious belief and practice, each person's location and date of birth put him on a cultural trajectory from which it is difficult to deviate with respect to insight or opinion.[5] Even when such departures do occur, it is almost always to become fixed on another equally invariant course. This is not entirely undesirable because a reasonably high level of stability is necessary for productive lives, but it can become problematic when one forgets that his current trajectory may have no sounder basis

---

[4]The remark (previously mentioned) that is sometimes heard in Christian circles—"God said it, I believe it, that settles it"—is presumably spoken by devout theists who see themselves taking God's Word as evidence for the rationality of their belief in the assertions of Scripture. While it certainly seems illogical to disbelieve the words of a god (and if God did say it one would probably do well to believe it), there is an obvious circularity in such reasoning, and deciding what God actually said (or says) can be a challenge all its own. It is not as simple as equating everything one reads in the Bible with a direct saying from God (e.g., 1 Cor 7:12, 25-40)—hence the need to be prepared to give a reason.

[5]Steve Donaldson, *Dimensions of Faith: Understanding Faith Through the Lens of Science and Religion* (Eugene, OR: Cascade, 2015), 121-25.

than the fact that it is the one into which he happened to be born or stumble. There is superficial acknowledgment of this influence among most literate individuals, but it is usually expressed in generalities that tend to obscure the fact that no one is immune—and that includes scientists as well as theologians.[6]

Recognizing the potential for a contextually formulated bias is a prerequisite to any sincere exploration for truth.[7] But even then it is possible to be deluded into believing that one has logically and freely chosen a particular path when in reality many of the bases for those choices are themselves beyond personal control.[8] Nevertheless, scientists who desire to integrate Christian faith and science will find that their scientific training plus the fruits of scientific understanding can perhaps contribute to their ability to identify and confront such a plight, both in themselves and in others.

Helping with interpretive issues. Many (if not all) of the problems at the conjunction of scientific and religious views arise because it has been predetermined that there is one privileged way to read Scripture, or one way to interpret scientific or historical evidence, or that what is meaningful is obvious and should be apparent to everyone. As noted above, what people often neglect is an admission that such determinations are not unbiased and that they are sometimes a reflection of what is happening to the person

---

[6]Cf. Stephen Wolfram, *A New Kind of Science* (Champaign, IL: Wolfram Media, 2002), 633.

[7]Cf. Francis Bacon's idols of the mind; *The Great Instauration*, in *The Works*, vol. 8, trans. James Spedding et al. (Boston: Taggard and Thompson, 1863).

[8]See Daniel Wegner, *The Illusion of Conscious Will* (Cambridge, MA: Bradford, 2003); Michael Gazzaniga, *Who's in Charge? Free Will and the Science of the Brain* (New York: HarperCollins, 2012).

more than a sign of rational reasoning. This neglect would be difficult to excuse were it not for the fact that it too is culturally conditioned. But recognizing the power of cultural biases can serve as a starting point for possible reconciliation or, at the very least, communication, and it is the beginning of any semblance of genuine control over and responsibility for one's beliefs.

There is, for example, a frequent tendency to confuse beliefs about the presumed literal characteristics or actions of God (patterned after scaled-up human attributes) with beliefs about the existence of God.[9] This confusion can lead to the literal claim that there is no God, but it can also distort what might otherwise be a clearer picture of God. Christian scientists who have wrestled with some of these issues might have an ideal opportunity to help others who are struggling with them.

Dealing with doubt. It is interesting to consider the idea that atheists and theists not only share mechanisms for how they arrive at their respective beliefs, but they frequently display comparable behaviors. Proponents of each side show signs of deep and fervent belief in their adopted perspective, although the actions of either can be a charade. People who appear to hold their beliefs with an iron grip may in fact be tiring from the effort or simply concealing withered hands behind a façade of invincibility. Much was made some years ago following the death of Mother Teresa about her private (at least until that time) "crisis

---

[9]Although many people can read biblical passages such as Isaiah 59:1 ("Surely the arm of the LORD is not too short to save, nor his ear too dull to hear") without thinking that God has literal arms and ears, they nevertheless act as though God is really just a glorified human. Thus everything from disaster to disease is sometimes attributed to God in the same way that we would attribute it to human agency.

of faith," particularly how someone apparently so devoted could entertain the doubts she did.[10] But for every Mother Teresa there is a public atheist who harbors analogous reservations about which his colleagues would be surprised to learn.

The real surprise would be to find that there actually are atheists or theists who have no doubts and have experienced no "crises of faith." Perhaps there are some, but unless one retreats into a protective shell, any exploration of the claims for and against God is sure to raise questions. C. S. Lewis once said, "Now that I'm a Christian I do have moods in which the whole thing looks very improbable: but when I was an atheist I had moods in which Christianity looked terribly probable."[11] It is likely that even the most devout atheist (or theist) will admit some probability for the existence (or lack thereof) of God, but that probability shouldn't be viewed as some mysterious number that magically appears and exists on its own. The probability reflecting any person's belief about the existence of God is really just an amalgam of myriad competing and contrasting beliefs about all sorts of potential evidence. No wonder it can fluctuate by the moment for atheist and theist alike.

> To think in any way that faith is the problem or that, more specifically, faith is a religious problem is to entirely misunderstand both faith and God.
>
> Steve Donaldson, *Dimensions of Faith*

---

[10]David Biema, "Her Agony," *Time* 170, no. 10 (2007): 36-43.
[11]C. S. Lewis, *Mere Christianity* (Westwood, NJ: Barbour, 1952), 119.

Christian scientists can play two key roles in this regard. First, they can help dispel the myth that Christianity is not based on evidence. Second, because much of their training has been involved with analyzing evidence they can help others learn to analyze claims about both Christianity and science (and their interaction). For any subject, doubt is almost always a response to concerns about evidence and rationality, and hence constitutes a rational component of faith. In fact, one could go so far as to say, "No doubt? No rationality!" In short, blind faith is not rational faith and there is no reason to think one form of it is better than another.[12]

Confronting slippery slope fears. We use this phrase to describe the belief that acknowledging a possible mistake with respect to a single issue will inevitably lead to eventual abandonment of one's entire philosophical, scientific or theological position. The individual plagued by this concern is thus locked into a static mindset, afraid that loosening his grip on what he currently believes will send him sliding straight into a relativistic (in the sense of the absence of absolutes) if not literal hell. This fear is usually expressed as, If I can't believe X, then what can I believe (from my current belief set)? The implication is that if X is false, there is no reason to think previous belief Y might not also be false. Y is seen as inextricably linked to X so that, for example, if Adam falls off the cliff he must necessarily drag Jesus with him. Although this might be the case for some beliefs, concluding that it must be so for all is an inductive fallacy—a logic error without warrant.

---

[12]See Donaldson, *Dimensions of Faith.*

It should be noted that scientists are subject to slippery slope fears no less than religious folks. Usually people contemplate this fear in terms of the perceived potential to lose their religious faith, but it works both ways. Consequently, acknowledging even the possibility that his esteem for science, causal logic and his own reasoning powers might require more serious scrutiny could have put Haldane on a slippery slope at the bottom of which his atheism might have been jarred loose.

This whole enterprise takes great discernment, particularly since there is always the danger of abandoning a true belief and also because there are plenty of individuals who would like nothing better than to exploit this fear in order to drive a wedge into an existing belief set with which they disagree. Nevertheless, being afraid to question a belief because of this fear is worse. Christian scientists must be especially attuned to these concerns and prepared to confront them as they arise both within themselves and among church members and prospective members.

Discovering truth. Truth has a nasty habit of being inconvenient. Faced with the truth one may also be faced with the need to diet, exercise, study, travel or otherwise change habits, friends or points of view. In extreme cases the search for truth can be accompanied by controversy or outright hostility. More often, it may simply take a backseat to concerns that an existing and valued (though not necessarily valuable) relationship will be lost. Sacrifice of social standing or peer approval accompanying admission that a previously held view may have been wrong is for some a greater nightmare than not having the truth.

It is easy to think this is primarily a problem for certain religious viewpoints, but, as Kuhn suggests, scientists may cling so tightly to their preferred paradigm that only death can tear it from their grasp.[13] Thus, while those steeped in the Judeo-Christian tradition are encouraged by the likes of Francis Collins to relax their grip on a literal reading of Genesis, biologists are asked by Stuart Kauffman to loosen their hold on "Natural Selection, which we might as well capitalize as though it were the new deity."[14] Both Collins and Kauffman, of course, advocate a change of view because they believe that doing so will promote greater understanding in their respective domains of interest and not because they are intent on creating untenable positions for those who take their advice. The issue here is not whether their specific suggestions are useful, but rather the potential similarity in responses by their respective audiences. Thus the scientist who is afraid that a consideration of self-organization might undermine the current status of natural selection responsible for the perceived order in nature shows traits uncomfortably reminiscent of the unflinching biblical literalist.

In the search for truth, the role of the Christian scientist is to help keep before others the image of a God who is capable of withstanding scrutiny and who "rewards those who earnestly

---

[13]Thomas S. Kuhn, *The Structure of Scientific Revolutions* (Chicago: University of Chicago Press, 1996); cf. Donaldson, *Dimensions of Faith*, 122. It is therefore noteworthy when a scientist publicly backpedals—for example, Bryce DeWitt's advocacy of Hugh Everett's many worlds interpretation of quantum mechanics, a position he had previously criticized; Peter Byrne, "The Many Worlds of Hugh Everett," *Scientific American* (December 2007): 98-105.

[14]Francis Collins, *The Language of God* (New York: Free Press, 2007); Stuart Kauffman, *At Home in the Universe: The Search for the Laws of Self-Organization and Complexity* (New York: Oxford University Press, 1996), 8.

seek him" (Heb 11:6). Of course, a god that cannot stand up to scrutiny does not deserve the designation any more than a god that can be fully scrutinized, but many individuals who assert God's omnipotence nevertheless act as though he has feet of clay that will crumble unless *they* protect him.[15] Although God is assumed to be with them in "the valley of the shadow of death" (Ps 23:4 NKJV) he is somehow unable to protect them in the shadow of scientific evidence or theological thinking they find distasteful. Protection, of course, may not be his intention, but despite proclaiming that he has their best interests at heart they cannot see how this is possible if they must question and possibly abandon some of their cherished ideas. Eventually, the atheist ends up marginalizing religious belief while the Christian marginalizes science, but only because both have marginalized God. Christian scientists, then, must continually remind themselves and those they interact with of the natural human tendency to tie God's hands with the cords of our limited imaginations.

Contemplating big questions. A limited view of the scientific enterprise only sees the immediate problems facing the scientist and fails to recognize the potential for science to contribute to a growing understanding of big questions—those questions of meaning and value that have been asked by countless individuals for significant periods of time. But consider the following questions:

- Where is the soul in a physical brain?

- What does modern neuroscience suggest about free will?

---

[15]See Donaldson, *Dimensions of Faith*, 217, on the potentially adverse result of trying to protect God.

- What does it mean to be human in an evolutionary context?

- Will transhumanist endeavors change our understanding of being made in God's image?

- How does meaning arise from mindless mechanisms?

- How does the apparent randomness seen in nature relate to God's providence?

For each of these questions (and many others) modern science provides insights that can help frame understanding and stimulate thinking. Christian scientists who see their work in this larger context—and particularly those who have taken the initiative to examine the relationship between their work and relevant theological and philosophical perspectives—can help their fellow church members and others see such questions in a new and potentially useful light.

Conclusion

Christian scientists are under obligation to live faithful lives at home, work and church no less than any other Christian, but the scientific training they have acquired also equips them to demonstrate how science can serve as a window into the nature and action of God in ways that can extend the vision of those whose expertise lies elsewhere. That training can also pay dividends when helping current and prospective church members bridge the perceived chasm between theories of modern science and claims of Christian faith—a gap that has been unnecessarily imposed by Christians and non-Christians alike who have failed to see the possibilities of integration.

FOR FURTHER READING

Barbour, Ian. *Religion and Science*. San Francisco: HarperOne, 1997.

Barfield, Owen. *Saving the Appearances: A Study in Idolatry*. Middletown, CT: Wesleyan University Press, 1988.

Barr, Stephen M. *Modern Physics and Ancient Faith*. South Bend: University of Notre Dame Press, 2014.

Brooke, John Hedley. *Science and Religion: Some Historical Perspectives*. Cambridge: Cambridge University Press, 2014.

Chesterton, G. K. *Orthodoxy*. New York: Barnes & Noble, 2007.

Collins, Francis S. *The Language of God: A Scientist Presents Evidence for Belief*. New York: Free Press, 2007.

Donaldson, Steve. *Dimensions of Faith: Understanding Faith Through the Lens of Science and Religion*. Eugene, OR: Cascade, 2015.

Giberson, Karl W. *The Wonder of the Universe: Hints of God in Our Fine-Tuned World*. Downers Grove, IL: InterVarsity Press, 2012.

Harrison, Peter. *The Territories of Science and Religion*. Chicago: University of Chicago Press, 2015.

Haught, John. *Science and Faith: A New Introduction*. Mahwah, NJ: Paulist Press, 2013.

McGrath, Alister E. *Science and Religion: A New Introduction*. 2nd ed. Malden, MA: Wiley-Blackwell, 2009.

Miller, Kenneth R. *Finding Darwin's God: A Scientist's Search for Common Ground Between God and Evolution*. New York: Harper Perennial, 2007.

Numbers, Ronald L., ed. *Galileo Goes to Jail and Other Myths About Science and Religion*. Cambridge, MA: Harvard University Press, 2010.

Polkinghorne, John. *One World: The Interaction of Science and Theology*. West Conshohocken, PA: Templeton Press, 2010.

———. *The Way the World Is: The Christian Perspective of a Scientist*. Louisville: Westminster John Knox, 2007.

Walton, John. *The Lost World of Genesis One: Ancient Cosmology and the Origins Debate*. Downers Grove, IL: IVP Academic, 2009.

Wright, N. T. *Scripture and the Authority of God: How to Read the Bible Today*. New York: HarperCollins, 2011.

NAME AND SUBJECT INDEX

SCRIPTURE INDEX

LITTLE BOOKS SERIES

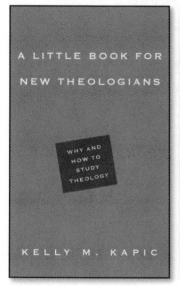

**A LITTLE BOOK FOR
NEW THEOLOGIANS**

By Kelly M. Kapic

**A LITTLE BOOK FOR
NEW PHILOSOPHERS**

By Paul Copan

Finding the Textbook You Need

The IVP Academic Textbook Selector
is an online tool for instantly finding the IVP books
suitable for over 250 courses across 24 disciplines.

---

www.ivpress.com/academic/
